AF780237

Application of Immunoassay Technology in Food Inspection

Application of Immunoassay Technology in Food Inspection

Editor

Maojun Jin

MDPI

Basel • Beijing • Wuhan • Barcelona • Belgrade • Novi Sad • Cluj • Manchester

Editor
Maojun Jin
Chinese Academy of
Agricultural Sciences (CAAS)
Beijing, China

Editorial Office
MDPI
St. Alban-Anlage 66
4052 Basel, Switzerland

This is a reprint of articles from the Special Issue published online in the open access journal *Foods* (ISSN 2304-8158) (available at: https://www.mdpi.com/journal/foods/special_issues/immunoassay_inspection).

For citation purposes, cite each article independently as indicated on the article page online and as indicated below:

Lastname, A.A.; Lastname, B.B. Article Title. *Journal Name* **Year**, *Volume Number*, Page Range.

ISBN 978-3-0365-8562-8 (Hbk)
ISBN 978-3-0365-8563-5 (PDF)
doi.org/10.3390/books978-3-0365-8563-5

© 2023 by the authors. Articles in this book are Open Access and distributed under the Creative Commons Attribution (CC BY) license. The book as a whole is distributed by MDPI under the terms and conditions of the Creative Commons Attribution-NonCommercial-NoDerivs (CC BY-NC-ND) license.

Contents

About the Editor

Maojun Jin

Maojun Jin is a Professor, doctoral supervisor, leading agricultural talent, and head of Research Division in the Agricultural Product Quality and Safety of Institute of Quality Standard and Testing Technology for Agro-Products, Chinese Academy of Agricultural Sciences (also named the Research Center for Agricultural Product Quality Standards of the Ministry of Agriculture and Rural Affairs). He received his bachelor's and doctoral degrees from Zhejiang University in 2004 and 2009, respectively. His research focus is the quality and safety of agricultural products. He is mainly engaged in the recognition theory and preparation of small-molecule compound antibodies based on monoclonal antibodies, nanobodies, recombinant antibodies, in research on the immunoassay method, and in theoretical research based on immunochromatography, fluorescence, chemiluminescence, and bio-barcode. He served as the Associate Editor of Frontiers in *Nutrition*, Co-editor in Chief of *Food Safety and Health*, Editorial Board member of SCI journals such as *Foods*, and also served as a member of the 10th National Pesticide Registration and Review Committee of the Ministry of Agriculture and Rural Affairs of China and as a specialist in a rapid testing standards review by the State Administration for Market Regulation of China.

Maojun Jin has led nearly 30 projects, including 3 National Natural Science Foundation projects, national key research and development plan projects, national major scientific instrument and equipment development special projects, international science and technology cooperation projects of the Ministry of Science and Technology, and national/agricultural industry standard formulation projects. More than 80 of his SCI papers have been publicly published in domestic and foreign journals, of which more than 30 have been published with him as the first author or corresponding author (with a maximum impact factor of 20.6). He has published two Chinese and English books as the chief editor or on the editorial committee; he has had six national invention patents authorized as the first complete person; he has released 65 national food safety standards for pesticide maximum residue limits as the first complete person; and he has won six academic awards, including the second prize of the State Technological Innovation Award, the first prize of the Beijing Science and Technology Award, and the first prize of the Science and Technology Award of the China Association for Analysis and Testing.

foods

Editorial

Application of Immunoassay Technology in Food Inspection

Peipei Li and Maojun Jin *

Institute of Quality Standard and Testing Technology for Agro-Products, Chinese Academy of Agricultural Sciences, Beijing 100081, China; 18331090835@163.com
* Correspondence: jinmaojun@caas.cn; Tel.: +86-10-8210-6570

Citation: Li, P.; Jin, M. Application of Immunoassay Technology in Food Inspection. *Foods* **2023**, *12*, 2923. https://doi.org/10.3390/foods12152923

Received: 25 July 2023
Accepted: 29 July 2023
Published: 31 July 2023

Copyright: © 2023 by the authors. Licensee MDPI, Basel, Switzerland. This article is an open access article distributed under the terms and conditions of the Creative Commons Attribution (CC BY) license (https://creativecommons.org/licenses/by/4.0/).

Food safety is as important as ever, and the safeguards implemented to inspect and reduce pesticides, veterinary drugs, toxins, pathogens, illegal additives, and other deleterious contaminants in our food supply has helped improve human health and increase the length and quality of our lives. Multiple countries and regions have established a series of food safety laws and national standards, which has driven the development of rapid detection methods for the measurement of hazards in food products. Benefiting from the merits of convenience, high efficiency, and on-site rapid detection, etc., immunoassays have achieved mainstream application in the rapid inspection field. As early as the mid-20th century, the first immunoassays were described for the measurement of insulin and thyroxine, respectively. Currently, hundreds of immunoassays have been established for scores of food hazards. Ouyang et al. [1] summarized the immunoassays used to detect β_2-Agonists, including enzyme-linked immunosorbent assays (ELISA), lateral flow immunoassay (LFIA), fluorescence immunoassay, chemiluminescence immunoassay, etc. Generally, immunoassays are based on the principle of the specific recognition of antigens and antibodies. Antibodies serve as the basic recognition elements for immunological detection methods. Therefore, the design of hapten with high immunogenicity and preparation of antibodies with high specificity and affinity has become the most important part of immunoassays. In addition, improving the detection mode of immunoassays has become a research hotspot in recent years. Thus, this Special Issue provides some general methods used for hapten design and for the preparation of recognition elements, as well as an overview of immunoassay technology in food inspection.

The hapten design is key to the establishment of an immunoassay. The natural toxins α-solanine (SO) and α-chaconine (CHA) comprise approximately 95% of the total glycoalkaloid (GA) content in potatoes; ingestion of more than 1 mg of GAs per kg of body weight is toxic to humans. In order to simultaneously measure SO and CHA, Okada et al. [2] used solanidine, a chemical compound found in both SO and CHA, as the hapten for rabbit immunity. Two new polyclonal antibodies (pAbs) that bind to both SO and CHA were purified and obtained. Xu et al. [3] designed and synthesized a novel hapten of triazine herbicide atrazine (ATR), namely, 2-chloro-4-ethylamino-6-isopropylamino-1,3,5-triazine. This hapten maximally retains the characteristic structure of ATR and induces expected antibody response in the animal's immune system. After cell fusion and screening, one monoclonal antibody (mAb) 9F5 was obtained with the isotype of IgG1. For the standard curve, the IC_{50} value of the 9F5 mAb was 1.678 μg/L with the working range (IC_{20}–IC_{80}) of 0.384 to 11.565 μg/L. In addition, this mAb showed high specificity to ATR and low cross-reactivity to other triazine herbicides. The prepared 9F5 mAb provided the core raw material for establishing an ATR immunoassay. In recent years, specific peptide ligands, such as peptidomimetic and anti-immunocomplex peptides, have been regarded as promising substitutes for chemical haptens. The peptidomimetic directly reacts with the antibody, while the anti-immunocomplex reacts with the immunocomplex between antibody and antigen. You et al. [4] identified thirty sequences of peptidomimetics and two sequences of anti-immunocomplex peptides for imidacloprid (IMI) from three phage display libraries, in which the anti-immunocomplex peptides were the first reported noncompetitive reagents for IMI. One peptidomimetic and one anti-immunocomplex peptide were utilized to develop competitive and non-competitive phage ELISA (P-ELISAs), with IC_{50} of 0.55 ng/mL and 0.35 ng/mL, respectively. Compared to chemical haptens, peptide ligands are prepared using biological expression. It is attractive to explore the peptide ligands to derive novel patterns of immunoassay.

In addition to the traditional pAbs and mAbs, more types of recognition elements have been discovered for immunoassay. Nanobodies, the smallest known functional antibody, have been used in the field of food safety. Qiu et al. [5] expressed a nanobody from a naive phage display library and used it as the capture antibody. Quantum dot (QB) was conjugated with a tnti-Cry2A pAb to serve as fluorescent probe. Then, a sensitive sandwich fluorescence-linked immunosorbent assay (QB-FLISA) using nanobody was established to determine the Cry2A toxin in cereal. The detection limit of this assay was 0.41 ng/mL, which had 19-times higher sensitivity than the traditional colorimetric ELISA. Known as "chemical antibodies", aptamer has the advantages of simple preparation, stable properties, low immunogenicity, and easy modification compared with the antibody. Zhang et al. [6] used aptamer as recognition element and developed a competitive enzyme-linked aptamer assay (ELAA) kit to detect Lactoferrin (Lf) in dairy products. In the construction, the Lf aptamer was conjugated with horseradish peroxidase (HRP) as the recognition probe and an aptamer complementary strand was anchored onto the microplate as the capture probe. When TMB substrate appeared in the reaction system, the color shades were negatively correlated with the Lf concentrations in the sample. Under the optimization conditions, the aptamer-based ELAA kit achieved a good linear relationship (R^2 = 0.9901) in the wide range of 25–500 nM with the detection limit of 14.01 nM and good specificity. In addition, Wang et al. [7] developed an aptamer-based surface-enhanced Raman spectroscopy (SERS) analysis for the anti-interference detection of Ochratoxin A (OTA). In this aptasensor, 4-[(Trimethylsilyl) ethynyl] aniline (TEAE) as an anti-interference Raman reporter, assembled on AuNPs with OTA-aptamer, served as SERS probes. Meanwhile, Fe_3O_4NPs, linked with complementary aptamer (cApts), were applied as capture probes. The specific binding of OTA to aptamer hindered the complementary binding of aptamer and cApt, resulting in a negative correlation between the Raman response at 1998 cm^{-1} and OTA levels. Under the optimum condition, the aptasensor presented a linear response for OTA detection in the range of 0.1–40 nM, with a low detection limit of 30 pM.

Many detection approaches have been developed to conduct qualitative/quantitative analysis of hazards in food samples. Zhu et al. [8] established an indirect competitive chemiluminescence enzyme-linked immunoassay (Cl-ELISA) to detect acetamiprid (ACE) in the Chinese cabbage and cucumber. The LOD of this assay was 1.26 μg/kg, meeting the MRL requirements of ACE in vegetables. Subsequently, Zhai et al. [9] compared Cl-ELISA with colorimetric ELISA (CO-ELISA) for IMI detection in vegetables. The results indicated that Cl-ELISA showed high sensitivity and a rapid detection time, saving costs (antigen and antibody concentrations) and serving as a more efficient model for the rapid detection of IMI residue, which provides a theoretical basis for selecting an optical detection assay. The improvement of the sensitivity of immunoassay has always been the focus of researchers. Hendrickson et al. [10] developed a cascade-enhanced LFIA to detect okadaic acid (OA) in seawater, fish, and seafood. In the case of the cascade-enhanced LFIA, the anti-OA antibodies specifically bound to the OA-BSA on the T line of the strip, followed by several (at least two) cycles of successive passing gold-labeled goat anti-mouse antibodies (GAMI–AuNPs) and free donkey anti-goat antibodies (DAGI) that were not specific to OA. As a result, branched aggregates formed on the T line with a (GAMI-AuNPs-DAGI) $\times n$ structure, where n is the number of GAMI-AuNPs/DAGI passing cycles. The LOD of the cascade-enhanced LFIA enabled the detection of OA at 30 pg/mL, which was two to seven times higher than the LOD of the LFIA without amplification. Immunomagnetic beads (IMBs) have been widely used to capture and isolate target pathogens from complex food samples. Kang et al. [11] proposed a pH-regulated strategy to orient the antibody on the surface of MBs. This study revealed that the positively charged ε-NH_2 group of lysine on the Fc relative to the uncharged amino terminus on Fab was preferentially adsorbed on the surface of MBs with a negatively charged group at pH 8.0, resulting in the antigen-binding sites of the antibody being fully exposed. This strategy can efficiently capture and isolate pathogenic microorganisms from complex food matrices at low cost, thereby improving the specificity and sensitivity combined with immunoassay.

In summary, this Special Issue provides insights into the recent application of immunoassay technology in food inspection. We sincerely hope that readers will find this Special Issue informative and interesting.

Author Contributions: P.L. and M.J. contributed equally to organizing the Special Issue, to the editorial work, and to writing this editorial. All authors have read and agreed to the published version of the manuscript.

Funding: This research received no external funding.

Conflicts of Interest: The authors declare no conflict of interest.

References

1. Ouyang, S.; Yu, S.; Le, Y. Current advances in immunoassays for the detection of β2-Agonists. *Foods* **2022**, *11*, 803. [CrossRef] [PubMed]
2. Okada, K.; Matsuo, K. Development of new antibodies and an ELISA system to detect the potato alkaloids α-Solanine and α-Chaconine. *Foods* **2023**, *12*, 1621. [CrossRef] [PubMed]
3. Xu, L.; Abd El-Aty, A.M.; Shim, J.H.; Eun, J.B.; Lei, X.; Zhao, J.; Zhang, X.; Cui, X.; She, Y.; Jin, F.; et al. Design and characterization of a novel hapten and preparation of monoclonal antibody for detecting atrazine. *Foods* **2022**, *11*, 1726. [CrossRef] [PubMed]
4. You, T.; Ding, Y.; Huang, Y.; Lu, Y.; Wang, M.; Hua, X. Identification and application of two promising peptide ligands for the immunodetection of imidacloprid residue. *Foods* **2022**, *11*, 3163. [CrossRef] [PubMed]
5. Qiu, Y.; You, A.; Fu, X.; Zhang, M.; Cui, H.; Zhang, B.; Qin, W.; Ye, Z.; Yu, X. Quantum-dot-bead-based fluorescence-linked immunosorbent assay for sensitive detection of Cry2A toxin in cereals using nanobodies. *Foods* **2022**, *11*, 2780. [CrossRef] [PubMed]
6. Zhang, F.; Du, H.; Li, L.; Li, T.; Wang, J.; Chen, Z.; Yan, M.; Zhu, C.; Qu, F. Enzyme-linked aptamer kits for rapid, visual, and sensitive determination of lactoferrin in dairy products. *Foods* **2022**, *11*, 3763. [CrossRef] [PubMed]
7. Wang, H.; Chen, L.; Li, M.; She, Y.; Zhu, C.; Yan, M. An alkyne-mediated SERS aptasensor for anti-interference ochratoxin a detection in real samples. *Foods* **2022**, *11*, 3407. [CrossRef] [PubMed]
8. Zhu, Z.; Shi, Q.; Wu, J.; He, K.; Feng, J.; Dong, S. Determination of acetamiprid residues in vegetables by indirect competitive chemiluminescence enzyme immunoassay. *Foods* **2022**, *11*, 2507. [CrossRef] [PubMed]
9. Zhai, R.; Chen, G.; Liu, G.; Huang, X.; Xu, X.; Li, L.; Zhang, Y.; Xu, D.; Abd El-Aty, A.M. Comparison of chemiluminescence enzyme immunoassay (Cl-ELISA) with colorimetric enzyme immunoassay (Co-ELISA) for imidacloprid detection in vegetables. *Foods* **2023**, *12*, 196. [CrossRef] [PubMed]
10. Hendrickson, O.D.; Zvereva, E.A.; Zherdev, A.V.; Dzantiev, B.B. Cascade-enhanced lateral flow immunoassay for sensitive detection of okadaic acid in seawater, fish, and seafood. *Foods* **2022**, *11*, 1691. [CrossRef] [PubMed]
11. Kang, F.; Yang, Y.; Li, J.; Chen, E.; Hong, T.; Zhao, L.; Du, M. pH-regulated strategy and mechanism of antibody orientation on magnetic beads for improving capture performance of staphylococcus species. *Foods* **2022**, *11*, 3599. [CrossRef] [PubMed]

Disclaimer/Publisher's Note: The statements, opinions and data contained in all publications are solely those of the individual author(s) and contributor(s) and not of MDPI and/or the editor(s). MDPI and/or the editor(s) disclaim responsibility for any injury to people or property resulting from any ideas, methods, instructions or products referred to in the content.

foods

MDPI

Review

Current Advances in Immunoassays for the Detection of β_2-Agonists

Shuyu Ouyang [1,†], Shuting Yu [1,†] and Yingying Le [1,2,*]

[1] CAS Key Laboratory of Nutrition, Metabolism and Food Safety, Shanghai Institute of Nutrition and Health, University of Chinese Academy of Sciences, Chinese Academy of Sciences, Shanghai 200031, China; ouyangshuyu2018@sibs.ac.cn (S.O.); yushuting2019@sibs.ac.cn (S.Y.)
[2] Key Laboratory of Food Safety Risk Assessment, Ministry of Health, Beijing 100021, China
* Correspondence: yyle@sibs.ac.cn
† These authors contributed equally to this work.

Abstract: β_2-agonists are a group of synthetic phenylethanolamine compounds which are traditionally used for treating bronchospasm. These compounds can also increase skeletal muscle mass and decrease body fat. The illegal use of β_2-agonists in food-producing animals results in residue of β_2-agonists in edible tissues and causes adverse health effects in humans. Thus, the detection of β_2-agonists at trace level in complex sample matrices is of great importance for monitoring the abuse of β_2-agonists. Many methods have been developed to detect β_2-agonists. Among them, a variety of antigen–antibody interaction-based techniques have been established to detect β_2-agonists in various samples, including animal feed, urine, serum, milk, tissues and hair. In this review, we summarized current achievement in the extraction of β_2-agonists from testing samples and detection of β_2-agonists using immunological techniques. Future perspectives were briefly discussed.

Keywords: β_2-agonist; sample extraction; immunoassay; radioimmunoassay; ELISA; chemiluminescent immunoassay; lateral flow immunoassay; immunosensor; food safety

Citation: Ouyang, S.; Yu, S.; Le, Y. Current Advances in Immunoassays for the Detection of β_2-Agonists. *Foods* **2022**, *11*, 803. https://doi.org/10.3390/foods11060803

Academic Editor: Maojun Jin

Received: 1 February 2022
Accepted: 4 March 2022
Published: 11 March 2022

Publisher's Note: MDPI stays neutral with regard to jurisdictional claims in published maps and institutional affiliations.

Copyright: © 2022 by the authors. Licensee MDPI, Basel, Switzerland. This article is an open access article distributed under the terms and conditions of the Creative Commons Attribution (CC BY) license (https://creativecommons.org/licenses/by/4.0/).

1. Introduction

β_2-agonists are a group of synthetic compounds with phenylethylamine structures, which are commonly used to treat asthma and chronic obstructive pulmonary disease [1]. They relax the smooth muscle of respiratory tract by combining to β_2 adrenoreceptor. β_2-agonists are also powerful anabolic agents which can promote protein synthesis, increase muscle mass and decrease fat tissue [2]. They are illegally used in food-producing animals as the growth promoters and nutrient repartitioning agents to escalate lean muscle gain, increase growth rate and feed efficiency [3–6]. Most countries around the world ban the use of all β_2-agonistsin livestock feed and have established strict surveillance programs to ensure the food and feed safety. However, the illegal use of β_2-agonists in livestock still happens, and the poisoning incidents caused by consumption of edible tissues from livestock bred with β_2-agonistsare reported from time to time in countries around the world [7–12].

In order to monitor the illegal use of β_2-agonists, various techniques have been developed to detect β_2-agonists in animal samples (tissues, milk, urine, hair, etc.), including chromatography, spectrometry and related techniques [13,14], immunoassays [13,14], biosensors [14,15] and β_2 adrenoreceptor-based assays [16,17]. Immunoassays are widely used in the purification and measurement of β_2-agonists. The antibodies against β_2-agonist can be prepared with β_2-agonist hapten composed of β_2-agonist and a carrier protein, such as serum albumin from bovine, human and rabbit, ovalbumin, keyhole limpet hemocyanin and bovine thyroglobulin. In this review, we summarize antigen–antibody interaction-based methods to purify and determine β_2-agonists, including extraction of β_2-agonists from samples through immunoaffinity chromatography, immunofiltration and immunomagnetic separation, and detection of β_2-agonists by radioimmunoassay, enzyme linked immunosorbent assay (ELISA), chemiluminescence immunoassay, lateral flow immunoassay, immunosensors and other types of immunoassays.

2. β_2-Agonist Antibody-Based Sample Extraction/Cleanup

Extraction and cleanup are important steps for the detection of β_2-agonists in complex biological samples. Various techniques have been developed to extract and cleanup β-agonists, such as liquid–liquid extraction, solid phase extraction, matrix solid-phase dispersion, dialysis, supercritical fluid extraction [13,14,18]; and antibody-based immunoaffinity chromatography [18–22], immunofiltration [23–25] and immunomagnetic separation [26–29]. Antibody-based techniques provide better cleanup of the samples and higher selectivity than aforementioned other techniques and were summarized herein.

2.1. Immunoaffinity Chromatography

Immunoaffinity chromatography (IAC) is a technique that relies on antigen–antibody interactions to extract the analyte(s) of interest. Analyte from the sample is retained on the column containing immobilized antibody and eluted using minimal amounts of organic solvent. IAC has been accepted as an extractionpreconcentration procedure for detecting β_2-agonists in biological samples owing to its high specificity and sample cleanup efficiency. IAC has been applied to extract clenbuterol, salbutamol, ractopamine and its metabolites from urine and tissue samples, respectively. Then, the target compound was detected using different techniques, including high-performance liquid chromatography (HPLC), electrochemical detection and capillary liquid chromatography-tandem mass spectrometry [19–21]. Lin et al. [22] developed a method to simultaneously detect clenbuterol, salbutamol, ractopamine and terbutaline in beef by IAC extraction and ultra-high-performance LC-MS/MS detection of these compounds. The immunoaffinity column was made by simultaneously covalent coupling of monoclonal antibodies against clenbuterol, salbutamol and ractopamine, respectively. As the antibodies are not specific for terbutaline, the limit of detection (LOD) of terbutaline is higher than that of the other three β_2-agonists.

2.2. Immunofiltration

Immunofiltration has been applied for sample cleanup for detecting β_2-agonists. The antibodies against β_2-agonist are mixed with the samples and incubated in an ultrafiltration device. After centrifugation, the filter is washed with buffer, and the antibody bound β_2-agonist is freed from the antibody by acetic acid. Immunofiltration was used to pretreat urine samples for detection of clenbuterol with a biosensor immunoassay [23] or ELISA [24]. Haasnoot et al. [25] reported that the anti-salbutamol polyclonal antibodies (pAb) recognized several β-agonists, and the combination of immunofiltration of β_2-agonists with the ELISA could detect at least ten β_2-agonists in urine with comparable LODs.

2.3. Immunomagnetic Separation

Immunomagnetic separation involves the coupling of biological macromolecules, such as antibodies and streptavidin, to superparamagnetic particles. When added to a heterogeneous target suspension, the magnetic particles bind to the desired target and form a complex which can be removed from the suspension by using a magnet. Immunomagnetic separation has been used as a sample pretreatment technology for purification and enrichment of β_2-agonists from samples. Chen et al. [26,27] prepared immunomagnetic beads using monoclonal antibodies against clenbuterol and salbutamol, respectively, purified these compounds from animal urine samples and detected them by surface-enhanced Raman spectroscopy. Peng et al. [28] separated clenbuterol from swine urine samples using immunomagnetic particles which was prepared by immobilizing biotinylated clenbuterol monoclonal antibody (mAb) on the streptavidin magnetic nanoparticles through biotin–streptavidin interaction. Huang et al. [29] modified the fluorescent magnetic nanobeads with anti-clenbuterol mAb to specifically recognize and capture clenbuterol in swine urine.

The fluorescent magnetic nanobeads not only served as a carrier for immunomagnetic separation of clenbuterol but also as a fluorescent probe for fluorescent lateral flow immunoassay of clenbuterol.

The advantages and disadvantages of the methods for extracting β_2-agonists from samples using antibodies are summarized in Table 1.

Table 1. Advantages and disadvantages of antibody-based methods for extracting β_2-agonists from samples.

Methods	Advantages	Disadvantages
Immunoaffinity chromatography	Simple, fast; high separation efficiency and reproducibility	Proteases from animal sample may digest protein A or antibody on the immunoaffinity column. The high-affinity binding of antibodies and antigens may complicate the elution of the antigen.
Immunofiltration	Simple, fast, low cost	Filters are easily clogged by large particles in samples. Contaminating proteins that adhered to the filter will elute with the antigens.
Immunomagnetic separation	Simple, fast, easy to perform; high separation efficiency and reproducibility	Commercial immunomagnetic particles for β_2-agonists extraction are not available.

3. Immunological Techniques for Detecting β_2-Agonists

Many immunological methods based on an antigen–antibody interaction have been developed for qualitative/quantitative analysis of β_2-agonists in biological samples, including radioimmunoassay, ELISA, chemiluminescence immunoassay, lateral flow immunoassay, immunosensors and other types of immunoassays.

3.1. Radioimmunoassay

The classical radioimmunoassay (RIA) is based on the principle of competitive binding. The unlabeled antigens and a fixed amount of radiolabeled antigens compete for a limited amount of antibody to form an antigen–antibody complex. The amount of labeled antigen–antibody complex formed is inversely related to the concentration of the unlabeled antigen. RIA has been applied to detect β_2-agonists, such as salbutamol, fenoterol and clenbuterol in various samples, including plasma/serum, urine, feces and liver tissue [30–34]. If the pAb against one β-agonist has cross-reactivity with other β-agonists, the RIA can be applied as a qualitative method to detect the presence of one or more β-agonists [33,34], then a confirmatory analysis would be performed to determine the concentration of the analyte(s). The use of RIAs in detecting β_2-agonists has been largely replaced by nonisotopic immunoassays because of concerns over the safe handling and disposal of radioactive reagents and waste.

3.2. ELISA

ELISA, also referred to as enzyme immunoassay, is a plated-based solid-phase enzyme immunoassay commonly used to identify the presence and concentration of antibodies or antigens in liquid samples. Detection of antigens is accomplished by binding enzyme–antibody conjugates to the antigen in the sample and assessing the enzyme activity via incubation with a substrate to produce a measurable end product. A number of enzymes are used in ELISA, including horseradish peroxidase (HRP), alkaline phosphatase (ALP) and β-galactosidase. ELISA can be classified into direct, indirect, sandwich and competitive

ELISA. The competitive ELISA is very useful for determining the concentration of small-molecule antigens in complex sample mixtures. Direct and indirect competitive ELISAs have been developed to detect β_2-agonist in various samples.

3.2.1. Direct Competitive ELISA

Clenbuterol [35,36], salbutamol [37] and phenylethanolamine A [38] in various samples (animal urine, blood, hair and tissues) have been measured using direct competitive ELISA. The mAb or pAb against β_2-agonist is coated onto the microtiter plate, the β-agonist in sample and β_2-agonist labeled with HRP or ALP competes for binding with the mobilized antibody. Addition of the substrate of the enzyme yields a signal that is inversely proportional to β-agonist concentration within the sample. Huang et al. [39] developed a direct competitive ELISA using bacteria-Au-antibody/HRP composite as a probe to detect clenbuterol. The clenbuterol in the sample competed with clenbuterol immobilized on a plate for binding to the limited probe. The use of bacteria as a carrier and Au nanoparticles as cross-linking agents enriched HRP and reduced the antibody applied, which resulted in the amplification of signal in detection. The liner range and limit of detection (LOD) of clenbuterol were 0.02–1.0 ng/mL and 0.03 ng/mL, respectively.

3.2.2. Indirect Competitive ELISA

Indirect competitive ELISAs have been developed to detect β_2-agonists, including phenylethanolamine A [40], ractopamine [41], zilpaterol [42], clenbuterol [43,44] and salbutamol [45,46] in various types of samples. In these assays, the β_2-agonist coated on the microtiter plate competes with β_2-agonist in samples for binding to the primary anti-β_2-agonist antibody. Probe labeled secondary antibody against the primary antibody was added successively. The signal produced by the secondary antibody is inversely proportional to β_2-agonist concentration within the sample. Most of the above indirect competitive ELISAs used HRP to label the secondary antibody. Fang et al. [46] used Eu^{3+}-labeled secondary antibody in the ELISA to detect salbutamol. Han et al. [41] developed a gold nanoparticle-based indirect competitive ELISA to detect ractopamine by using a secondary antibody labeled with catalase through biotin–streptavidin interaction. In the presence of ractopamine in the sample, gold(III) ions are oxidized by H_2O_2 to form red AuNPs. In the absence of ractopamine, the AuNPs in the solution are purple or blue due to aggregation. The LOD for ractopamine in urine was 0.35 ng/mL. In general, the primary antibody specifically against one β_2-agonist was used in the development of the indirect competitive ELISA. Wang et al. [44,45] developed indirect competitive ELISAs using polyclonal R-(-)-salbutamol antibody which could recognize 31 β-agonists and analogues, and clenbuterol monoclonal antibody which could recognize 23 β-agonists and analogues, respectively. Holographic and three-dimensional quantitative structure–activity relationship models were developed for predicting the epitopes on *β*-agonist hapten affecting antibody specificity, which will contribute to the rational design and control of the immunoassay specificity of β_2-agonist.

3.3. *Chemiluminescence Immunoassay*

Chemiluminescence describes the emission of light that occurs as a result of unique chemical reactions. In a chemiluminescence reaction, energy is released in the form of photons when electronically excited molecules, produced as a result of the reaction, relax to a stable ground state. Chemiluminescence immunoassay (CLIA) combines chemiluminescence technique with immunochemical reactions using luminescent chemical which could generate light emission. In CLIA, the luminescent chemical is used to label antigen, antibody or used as substrate of enzymes. HRP and ALP are the most widely used enzymes. Luminol and AMPPD are common chemiluminescent substrates used for HRP and AP, respectively.

CLIAs have been developed based on competitive immunoassay to detect salbutamol and brombuterol, respectively. HRP-labeled secondary antibodies were introduced into

CLIAs [47,48]. The chemiluminescence intensity was linearly related to the concentration of salbutamol in the range of 0.5–100 ng/mL with a LOD of 0.15 ng/mL [47]. The LOD of brombuterol was 0.33 pg/mL [48]. Wang et al. [49] developed a competitive chemiluminescence immunosensor based on a microfluidic chip to detect ractopamine in swine urine samples using HRP-labeled secondary antibody. The immunosensor could provide a liner range of 0.5–40 ng/mL with a LOD of 0.97 ng/mL for ractopamine detection.

3.4. Lateral Flow Immunoassay

Lateral flow immunoassay (LFIA), also known as immunochromatographic assay, is a combination of chromatography and immunoassay using a nitrocellulose membrane as a support to detect labeled antigen–antibody complexes in liquid samples. The signal produced on the membrane can be evaluated by the naked eye or with the aid of portable devices. Direct assay (sandwich assay) and competitive assay are two standard formats of LFIA, which are used to detect large and small molecules, respectively. Competitive tests have been used to detect β_2-agonists. A variety of tracers that cause color or optical changes following antigen–antibody interactions have been used in LFIA to detect β_2- agonists, such as nanoparticles, fluorescent nanomaterials and upconverted phosphorus nanoparticles. Compared with traditional detection methods, LFIA possesses significant advantages, including simple fabrication, low cost, fast and simple analytical procedure.

3.4.1. Colorimetric LFIA

Nanoparticles with specific structures can produce color by assemblies and aggregations. Various nanoparticles with specific properties have been used as tracers for LFIA to detect β_2-agonists in samples.

Gold Nanoparticles as Tracers

Colloidal gold or gold nanoparticle is the most widely used label in LFIA for detecting β_2-agonists in samples. It has an intense color and no development process is needed for visualization. Colloidal gold has been used by multiple laboratories to label an antibody against β_2-agonist and develop LFIA for detecting target β_2-agonist in biological samples, such as clenbuterol in urine and pork muscle [50,51], salbutamol in urine, meat and milk [52,53], phenylethanolamine A in urine and pork [40,54] and ractopamine in animal urine, meat, liver and feed [50,55–58]. Huang et al. [59] used bacteria as a carrier of gold nanoparticles to construct bacteria-Au-antibody probe in LFIA to detect clenbuterol. Much fewer antibodies are needed in the probe to produce a clearly visible color. The visual detection limit (VDL) for clenbuterol is 0.1 ng/mL in urine, 0.5 ng/mL in milk and 0.2 ng/g in swine feed. Wang et al. [60] developed silver–gold nanoparticles (Au-Ag NPs) probe-based LFIA to detect clenbuterol. The quantitative detection limit of optimized hollow Ag-Au NPs labeled lateral-flow immunochromatographic test strip approaches to 2 ppb. Compared with Au NPs and Ag NPs-labeled test strips, Ag-Au hollow NPs-labeled test strips exhibited much higher sensitivity for qualitatively detecting clenbuterol. Chen et al. [61] established an integrated LFIA to detect clenbuterol using anti-clenbuterol antibody labeled with colloidal gold and fluorescent nanobeads, respectively. The integrated test strip could qualitatively and quantitatively detect clenbuterol with VDL of 0.5 ng/mL and LOD of 0.04 ng/mL.

Other Nanoparticles as Tracers

Colored silica nanoparticles (SiNPs) have been used as a visible label for LFIA to detect β_2-agonists. Zhu et al. [62] developed LFIA to detect clenbuterol using purple SiNPs labeled anti-clenbuterol antibody. The VDL for clenbuterol were 6 ng/mL and 5 ng/mL in urine and pork, respectively. Yu et al. [63] developed colored SiNP-based LFIA to detect clenbuterol and ractopamine simultaneously by using anti-clenbuterol antibody and anti-ractopamine antibody labeled with red and blue SiNPs, respectively. The visible limit of detections for clenbuterol and ractopamine were 3 ng/mL and 2 ng/mL, respectively. Wang et al. [64,65] used selenium nanoparticles to label antibodies against clenbuterol, ractopamine and salbutamol, respectively, and developed LFIAs to detect these compounds in swine urine. Zhao et al. [66] developed LFIA for clenbuterol detection using Prussian blue nanoparticles-labeled antibody against clenbuterol. The VLDs for clenbuterol were 3.0 ng/mL in pork, 5 ng/g in both swine kidney and bacon samples. Liu et al. [67] used ultramarine blue nanoparticles as visible labels in LFIAs to detect ractopamine. The visual limit of detection for ractopamine is 2.0 ng/mL, and 1.0 ng/mL in feed and pork samples, respectively.

3.4.2. Luminescent LFIA

Luminescent materials, such as fluorescent nanomaterials and up-converting phosphorus nanoparticles, have been used as reporter to develop sensitive LFIA for detecting β_2-agonists.

Various kinds of fluorescent nanoparticles have been used as probes to develop LFIA for detecting β_2-agonists. Song et al. [68] used fluorescent nanosilica conjugated mAb against clenbuterol as a signal probe to build a LFIA for detecting clenbuterol. The VLD for qualitative detection was 0.1ng/mL and the LOD for quantitative detection was 0.037 ng/mL, which was much better than the colloidal gold-based strip. Huang et al. [29] developed an LFIA to detect clenbuterol using fluorescent magnetic nanobeads (FMNBs) to label anti-clenbuterol mAb. The FMNBs-Ab probe worked as a carrier for immunomagnetic separation of clenbuterol from sample and as a fluorescent label for the LFIA. The LOD of clenbuterol in swine urine is 0.22 ng/mL. Wang et al. [69] developed fluorescent beads-based LFIA to simultaneously detect 3 β_2-agonists. Fluorescent beads were conjugated with mAb specific for clenbuterol, ractopamine and salbuterol, respectively, and put on the conjugated pad. There were three test lines in the analytical pad, which contained immobilized clenbuterol, ractopamine and salbuterol, respectively. The LODs for clenbuterol, ractopamine and salbuterol were 0.10, 0.10 and 0.09 ng/mL, respectively, which were better than that of the colloidal gold-based strip. The same lab developed an LFIA to detect clenbuterol and its structural analogues in pork, using fluorescent nanoparticle-conjugated anti-clenbuterol antibody which had cross-reactivity with mabuterol, brombuterol, cimaterol, cimbuterol, bromchlorbuterol and banbuterol [70]. The established LFIA could screen these seven β_2-agonists in a single run. The LODs for these compounds in pork were <50 pg/g. Hu et al. [71] compared time-resolved fluorescent nanobeads, fluorescent submicrospheres, quantum dots and colloidal gold-based LFIA (TRFN-LFIA, FM-LFIA, QD-LFIA, and CG-LFIA) for detection of ractopamine in swine urine. TRFN-LFIA showed the highest sensitivity (LOD 7.2 pg/mL) and a wide linear range of detection (5–2500 pg/mL). In addition, TRFN-LFIA exhibited the shortest detection time compared with the other LFIAs. Shi et al. [72] developed a fluorescence quenching-based LFIA for detecting ractopamine. The antibody against ractopamine was labeled with gold nanoparticles (AuNPs) and loaded on the conjugate pad. The test line was coated with Rac and fluorescent polymer dots (FPDs). The control line was coated with FPD. Adding a negative sample resulted in the binding of Ab-AuNPs to ractopamine immobilized on T-Line, which in turn caused fluorescence resonance energy transfer between the FPDs and AuNPs and quenches fluorescence. Adding a positive sample caused a reaction between the ractopamine analyte and Ab-AuNPs and visible fluorescence at the test line. The intensity of fluorescence at the test line positively reflected the amount of ractopamine in the sample. The LOD was 0.16 ng/mL.

Wang et al. [73] developed an LFIA based on upconversion phosphor (UCP) for detection of clenbuterol. The mAb against clenbuterol was labeled with UCP beads which can emit higher energy light under excitation of lower energy light. After reaction, luminescence from clenbuterol-mAb-UCPs can be detected by a scanning photometer or naked eyes. The visual limit of detection for clenbuterol was 0.1 ng/mL. An LFIA based on a time-resolved chemiluminescence was established to detect ractopamine and clenbuterol simultaneously [74]. Antibodies against ractopamine and clenbuterol were conjugated with

HRP and ALP, respectively. As the reaction kinetics of HRP and ALP chemiluminescent reaction systems were different, ractopamine and clenbuterol could be sequentially detected in different time windows. The LODs for ractopamine and clenbuterol were 0.17 ng/mL and 0.067 ng/mL, respectively.

3.4.3. Other Types of LFIA

Zhang et al. [75] developed a competitive LFIA to detect clenbuterol in milk, swine liver and tenderloin using Coomassie Brilliant Blue (CBB)-stained anti-clenbuterol antibody. The CBB-antibody was used as both a recognition reagent and a chromogenic probe, enabling the simple but sensitive detection of clenbuterol with LOD of 2 ng/mL.

Surface-enhanced Raman scattering (SERS)-based LFIA has been developed for the detection of β_2-agonists, such as phenylethanolamine A [76], clenbuterol [77] and brombuterol [78]. The principle of this method is similar to the LFIA based on colloidal gold particles, but the anti-β_2-agonist antibody was labeled with Au@Ag core–shell nanoparticles sandwiched with a Raman reporter (4-mercaptobenzoic acid, MBA) [76,77], or with flower-like gold–silver core–shell bimetallic nanoparticles carrying the MBA [78]. The LOD values for detection of phenylethanolamine A, clenbuterol and brombuterol were 0.32, 0.24 and 0.5 pg/mL, respectively [76–78].

3.5. *Immunosensors*

Immunosensors use antibodies as the recognition element and a transducer that converts the antigen–antibody binding event to a measurable physical signal. Surface plasmon resonance, surface-enhanced Raman scattering, electrochemiluminescence and electrochemical immunosensor are types of immunosensors developed for detecting β_2-agonists.

3.5.1. Surface Plasmon Resonance Sensor

Surface plasmon resonance (SPR) is a phenomenon occurring at metal surfaces when plane-polarized light hits a metal film under total internal reflection conditions. SPR signal is directly dependent on the refractive index of the medium on the metal film. The binding of biomolecules on the metal surface results in changes in the refractive index. As SPR allows real-time, label-free detection of biomolecular interactions, it has been used as a powerful tool to study interactions between biomolecules, such as the interaction of antibody and antigen, ligand and receptor, enzyme and substrate [79].

A number of investigators developed competitive immunoassays to detect β_2-agonists using SPR sensors. A known concentration of β_2-agonist antibody was mixed with the sample and injected over the surface of the sensor chip. The β_2-agonist in the sample bound to the antibody and inhibited it from binding to the β_2-agonist mobilized on the sensor chip surface and consequently induced an increase in the refractive index at the SPR sensor surface. SPR sensor has been built to detect salbutamol in swine urine with LOD of 5 ng/mL [80]. Suherman et al. [81] fabricated SPR sensors with covalently immobilized β-albuterolvistion on Au chips for detection of ractopamine and salbutamol, respectively. The LODs for ractopamine and salbutamol were 10 pg/mL and 5 pg/mL, respectively. The sensor surface could be regenerated and reused more than 100 times. Wang et al. [82] developed a SPR to detect ractopamine in swine urine through indirect competitive assay, with an LOD of 0.09 ng/mL. Kabiraz et al. [83,84] built SPRs sensors to detect clenbuterol using Au nanoparticles to label the primary antibody against clenbuterol or the secondary antibody. The LODs were 0.05 pg/mL and 100 fg/mL, respectively, which were lower than that of unlabeled Ab.

3.5.2. Surface-Enhanced Raman Scattering-Based Immunosensor

Surface-enhanced Raman scattering (SERS) is a powerful spectroscopy technology that can provide a nondestructive and ultra-sensitive characterization down to single molecular level. Competitive SERS immunoassay has been applied to detect β_2-agonists, based on the competition between free β_2-agonists in samples and β_2-agonists immobilized on the solid substrate for binding to antibodies on SERS nanoprobes. The intensity of SERS signal is conversely correlated with the concentration of β_2-agonist in sample. The SERS nanoprobe is prepared by labeling Au nanoparticles (AuNPs) with Raman reporter and anti-β_2-agonist antibody. 4,40-dipyridyl and 2,20-dipyridyland 4-mercaptobenzoic acid (MBA) have been used as Raman reporters to prepare SERS nanoprobes for detecting clenbuterol and ractopamine, respectively [85,86]. Gu et al. [87] developed a SERS-based competitive immunoassay for detecting salbutamol and brombuterol in swine meat, liver and human urine. The SERS probe was prepared by immobilizing antibody against salbutamol (or brombuterol) and SERS reporter nile blue (or 3,3′,5,5′ -tetramethylbenzidine)-labeled DNA concatemers on gold nanoparticles. The introduction of long DNA concatemers in SERS probe and stain of the immune complex with Ag nanoparticles after competitive immunoreaction greatly enhanced Raman signals. The LODs for salbutamol and brombuterol were 2.0 pg/mL and 1.0 pg/mL, respectively. Wei et al. [88] developed a SERS-based liquid magnetic competitive immunoassay to detect clenbuterol. The SERS probe was prepared by labeling AuNPs with Raman reporter MBA and clenbuterol antibody. Fe_3O_4@Au nanoparticles were labeled with clenbuterol. After competitive reaction, the immunocomplex of SERS probe and Fe_3O_4@Au-clenbuterol were enriched by an external magnetic field and the SERS signal was collected. The LOD for clenbuterol was 0.22 fg/mL. Yao et al. [89] developed a SERS/resonance Rayleigh scattering (RRS) dual-spectroscopic immunosensor to detect clenbuterol based on nitrogen/silver-codoped carbon dots (CDN/Ag) catalytic amplification. CDN/Ag can strongly catalyze trisodium citrate-$HAuC_4$ reaction to generate red nanogold that exhibits a strong RRS signal. After adding Victoria blue B (VBB), the system also exhibits a strong Raman signal. Electrostatic coupling between CDN/Ag and Ab inhibits the catalytic activity of CDN/Ag and reduces the Raman signal. In the presence of clenbuterol, the binding of Ab with clenbuterol reduces the interaction of Ab with CDN/Ag and restore the catalytic activity of CDN/Ag, which enhances the Raman signal through increasing gold nanoparticles production. The SERS/RRS immunosensor detected clenbuterol with an LOD of 0.68 pg/mL.

3.5.3. Electrochemiluminescence Immunosensor

Electrochemiluminescence (ECL) is the process whereby species generated at the electrode undergo a high-energy electron transfer reaction to form excited states that emit light. ECL immunosensor combines immunoreaction and ECL, which can be used to quantitatively measure an antigen or antibody based on the change in ECL signal before and after immunoreaction. The ECL luminophore is one of the most significant components during the light generation processes. Various ECL luminophores have been reported, including organic, metal complexes, nanomaterials, metal oxides and near-infrared ECL luminophores [90]. Nanomaterials, such as quantum dots (QDs), have been adopted as final emission species either after direct oxidation reactions or after chemiluminescence resonance energy transfer.

Cadmium selenide (CdSe) QDs are one of the most popular ECL emitters due to their intrinsic properties, such as unique luminescent properties and relatively low cost. CdSe QDs have been used in ECL immunosensor to detect β_2-agonists. CdSe QDs immobilized on an electrode surface could react with H_2O_2 to produce a high ECL emission. In the presence of HRP substrate, the consumption of H_2O_2 in the HRP-catalyzed oxidation process leads to quenching of ECL emission. HRP is conjugated to anti-β_2-agonist Ab or the secondary Ab. In the competitive ECL immunoassay, β_2-agonist in the sample competes with β_2-agonist immobilized on the electrode for the primary Ab and reduces the amount of HRP-containing immunocomplex on the electrode, which alleviates the quenching of the ECL emission by reducing the consumption of H_2O_2 by HRP. Thus, the ECL intensity was proportional to the amount of β_2-agonist in sample. ECL immunosensors based on CdSe QDs and HRP-labeled secondary Ab have been constructed to detect salbutamol in pork and liver, with an LOD of 8.4 pg/mL [91]. ECL immunosensors based on CdSe QDs and anti-salbutamol or anti-ractopamine Ab-conjugated AuNPs-HRP have been constructed to detect these compounds in pork, liver and feed. AuNPs provided a matrix to anchor a large amount of HRP which greatly enhanced the electrochemical quenching. LODs for salbutamol and ractopamine were 17 pg/mL and 1.7 pg/mL, respectively [92,93].

To construct ECL immunosensors to detect β_2-agonist, many researchers used QDs as probe to label the anti-β_2-agonist antibody and used AuNPs as substrates and electron transport accelerators. By competitive immunoassay, the β_2-agonist in the sample competes with the β_2-agonist immobilized on the electrode for binding to QDs labeled β_2-agonist Ab, which results in the reduction in the β_2-agonist-Ab-QDs complex on electrode and ECL signal. Dong et al. [94] used CdSe@SiO_2 nanoparticles to label antibody against salbutamol to construct an ECL immunosensor to detect salbutamol in pork. The detection range was 0.001–1000 pg/mL with an LOD of 0.17 pg/mL. To amplify the ECL signal, various chemicals which can carry more QDs have been used to prepare the antibody probe for detecting β_2-agonists. Dong at al. [95] used polyamidoamine dendrimers (PAMAM) and silver-cysteine hybrid nanoribbon (SNR) to prepare CdSe QDS-PAMAM-SNR-Ab for detecting brombuterol in pork and feed. Zhu et al. used CdTeQDs-PAMAM-GO-labeled antibody composite to detect brombuterol [96] and used CdSe QDs/PDDA-GN/AuNPs-labeled antibody composite to detect ractopamine in pork [97]. The amplification of the ECL signal increased the sensitivity of these ECL immunosensors and decreased the LOD for β_2-agonist to lower pg/mL level.

Li et al. [98] developed an ECL resonance energy transfer (RET) immunosensor to detect brombuterol residues in pork and swine feed based on competitive immunoassay. The electrode was modified with graphitic-phase carbon nitride (g-C_3N_4) which worked as the ECL emitter. The antibody against brombuterol was labeled with Au-Ag alloy nanoparticles which were energy acceptors. When there is no brombuterol in the sample, the Au-Ag antibody binds with brombuterol immobilized on g-C_3N_4-coated electrode. Resonance energy transfer occurs and the ECL emission was quenched. On the contrary, brombuterol in the sample competed with brombuterol immobilized on g-C_3N_4 to bind Au-Ag antibody, g-C_3N_4 releases a strong ECL signal. The ECL-RET immunosensor for brombuterol detection exhibited high sensitivity with an LOD of 0.31 pg/mL.

3.5.4. Electrochemical Immunosensor

Electrochemical immunosensors combine highly sensitive electrochemical sensing technology with highly specific immunological techniques. In electrochemical immunosensor, the antibody and antigen interaction results in the change in electrical signal, such as current, voltage, resistivity and impedance, that correlates with the concentration of analyte.

Label-Free Electrochemical Immunosensor

Highly sensitive label-free electrochemical immunosensors have been constructed by immobilizing anti-β_2-agonist antibody on various types of electrodes which carry a gold nanostructure and other modifiers for detecting β_2-agonist, such as salbutamol [99], ractopamine [100] and clenbuterol [101].The LOD was at pg/mL or fg/mL level. Cui et al. [102] constructed a sandwich-type electrochemical immunosensor for detecting salbutamol. The primary salbutamol antibody was immobilized onto sodium dodecylbenzene sulfonate-functionalized graphene sheets (SDBS-GS) and coated on the electrode. The secondary salbutamol antibody was conjugated to Pd@SBA-15/BMIM·Br nanoparticles. The electrochemical immunosensor exhibited a wide working range from 0.02 to 15.0 ng/mL with an LOD of 7 pg/mL. Wang et al. [103] constructed a multiplexed electrochemical biosensor to simultaneously detect ractopamine (RAC), salbutamol (SAL) and clenbuterol (CLB) through a competitive immunoassay. Artificial antigens of RAC, SAL and CLB were, respectively, immobilized onto three working electrodes by binding to reduced graphene oxide. The antibody against RAC, SAL or CLB was conjugated with silver-palladium alloy nanoparticles (AgPd NPs). β_2-agonists in the sample competed with RAC, SAL and CLB on the electrodes for binding antibodies on AgPd NPs. This immunosensor can simultaneously detect RAC, SAL and CLB ranging from 0.01 to100 ng/mL with LOD of

1.52, 1.44 and 1.38 pg/mL, respectively. Gu et al. [104] developed three electrochemical immunosensors. Each can detect six β2-agonists through a competitive immunoassay. β_2-agonists in the sample competed with ractopamine (RAC), clenbuterol (CL) or salbutamol (SAL) immobilized on graphene-coated glassy carbon electrode (GCE/GNP) to bind with pAb against BSA-RAC-CL-SAL antigen. As the antibody had cross-reactivity with terbutaline, mabuterol and tulobuterol, each immunosensor could detect six β_2-agonists. All three immunosensors could detect CL with LODs of 0.1 ng/mL. The GCE/GNP/RAC immunosensor detected the other five β_2-agonists with LOD of 0.1 ng/mL, which was lower than those of the other two immunosensors.

Enzyme-Labeled Electrochemical Immunosensors

Enzymatic-labeled electrochemical immunosensors have been constructed to detect β_2-agonist based on competitive immunoassay, using electrode coated with β_2-agonist or primary β_2-agonist antibody/secondary antibody, and enzyme-labeled β_2-agonist antibody or β_2-agonist. The most commonly used enzymes are HRP and ALP. To enhance the electroactivity, the electrodes are usually modified with various materials, such as gold nanoparticles, polyaniline/poly(acrylic acid), colloidal Prussian blue and multiwall carbon nanotubes. Various nanomaterials are used as carriers of enzyme-conjugated β_2-agonist/antibody to amplify the signal for β_2-agonist detection.

Electrochemical immunosensors were developed to detect salbutamol by immobilizing salbutamol on the electrode modified with gold nanoparticles or other materials, and conjugating the HRP antibody to graphene, Au hybrid graphene nanocomposite or multiwall carbon nanotubes [105–107]. The LODs were between 0.03 ng/mL and 0.06 ng/mL, respectively. Electrochemical immunosensors have been fabricated to detect clenbuterol by immobilizing a primary or secondary antibody against clenbuterol on the electrode with various modifications, and labeling clenbuterol with HRP, ALP or glucose oxidase [108–111]. The LOD of clenbuterol is between 0.008 and 0.25 ng/mL.

3.6. Other Types of Immunoassays

In addition to the aforementioned immunological techniques, other methods based on the antigen–antibody interaction were also established for detecting β_2-agonists, such as immune-PCR, gel-based immunoassay, fluorescence polarization immunoassay and integrated microfluidic immunoassay.

Immuno-PCR is a technique to detect antigens and antibodies through a combination of an immunoassay and a PCR. In 1992, Sano et al. [112] developed an immuno-PCR technique based on the direct ELISA, in which the detection enzyme in ELISA was replaced with a biotinylated reporter DNA bound to the antigen–antibody complex through a streptavidin-protein A fusion protein. It combined the antigen detection with ultrasensitive PCR signal amplification, which greatly increases the sensitivity of detection. In recent years, immuno-PCR techniques have been developed by combining versatile DNA-antibody linking methods with real-time PCR (qPCR) to detect β_2-agonists. Lei et al. [113] established a direct competitive immuno-PCR assay to detect salbutamol in human urine. Each PCR tube was coated with salbutamol antibody, to which salbutamol labeled with double-stranded DNA and free salbutamol in the sample are competitively bound. The bound DNA in tubes was amplified and quantified by qPCR. Under the optimized conditions, the assay showed a linear range over seven orders of magnitude, whereas the LOD of salbutamol in the human urine sample was 28 fg/mL. Zhao et al. [114] developed a complete antigen-bridged DNA strand displacement amplification immuno-PCR assay for detecting salbutamol. Salbutamol antibodies were immobilized on the immunomagnetic beads. Salbutamol-BSA was linked to DNA probe 1 or 2 which each contained a split DNA barcode and a recognition site for cleavage endonuclease and subsequent quantification by qPCR. Upon binding with salbutamol antibodies on the immunomagnetic beads, the salbutamol probe 1 and probe 2 were closed and form a full-length DNA amplicon after the addition of a complementary bridging DNA and ligase. The full-length DNA amplicons were separated by magnetic separation and amplified by strand displacement amplification (SDA) and qPCR. To detect salbutamol, the salbutamol in the sample competes with salbutamol probes for binding salbutamol antibody, resulting in the reduction in full-length DNA amplicons necessary to SDA and qPCR. The developed immuno-PCR assay specifically detected salbutamol with LOD of 0.65 fg/mL. It was successfully applied to the detection of salbutamol in water and urine samples.

Li et al. [115] developed a gel-based immunoassay for detecting salbutamol and ractopamine residues in pork. The anti-salbutamol antibody and anti-ractopamine antibody were immobilized onto the Sepharose 4B gel, respectively. The anti-salbutamol gel and anti-ractopamine gel were loaded into 1 mL Bond Elut cartridges as the test layers, separated by a layer of air. The target analytes in the samples would compete with HRP-analyte conjugates to combine with the antibody in the test layer. After adding the chromogenic substrate, the negative samples presented a blue color on the respective test layers, while the positive samples presented no or negligible color. The color of the test layers was visually evaluated or photographed for color density-based quantitative analysis. The LODs were 0.5 μg/kg for salbutamol and ractopamine by visual detection. The quantitative LODs of salbutamol and ractopamine in spiked pork samples were 0.051 μg/kg and 0.02 μg/kg, respectively.

Fluorescence polarization immunoassay (FPIA) is based on measuring the polarization of light caused by changes in molecular size as a result of antigen–antibody reactions. FPIA use a fluorescein-labeled antigen to compete with the unlabeled antigen for an antibody. If the fluorescein-labeled antigen combines with the antibody, emitted light remains polarized when the incoming light is polarized. As the concentrations of antigen in the sample increase, free fluorescein-labeled antigen molecules also increase in number. Since the unbound fluorescein-labeled antigen molecule rotates freely, the polarized light emitted is reduced. FPIA has been widely applied to small molecule analysis. Zvereva et al. [116] detected ractopamine through FPIA, using ractopamine-aminomethyl fluorescein conjugate and mAb against ractopamine. The FPIA demonstrated an LOD of 1 ng/mL, range of detectable concentrations 2.3–50 ng/mL, and high specificity. It can be effectively used to test meat products. Dong et al. [117] developed FPIA for the detection of RAC in pork by synthesizing 10 fluorescein-labeled ractopamine derivatives (tracers) and paring with two pAbs against ractopamine. After careful selection and comparison, a highly sensitive and specific FPIA was developed with an LOD of 0.56 μg/kg for ractopamine in pork.

Zuo et al. [118] developed a method to simultaneously detect clenbuterol (CLB), ractopamine (RAC) and salbutamol (SAL) by hapten microarray-based indirect competitive immunoassay. The BSA conjugates of CLB, RAC and SAL were immobilized on slides precoated with agarose. The corresponding mAbs against these β-agonists, and the standards or samples were introduced for indirect competitive immunoassay. The antigen–antibody binding on the slide was detected using Cy3 dye-labeled secondary antibody. The LODs of CLB, RAC and SAL were 0.09, 0.50 and 0.01 μg/L, respectively. The hapten microarray system can perform high throughput and parallel analysis with high sensitivity and selectivity.

Kong et al. [119] established an integrated microfluidic immunoassay system for high throughput detection of clenbuterol. The 3-layer microchip had eight analysis units in which sample injection, washing, immunoreaction and enzyme catalysis were performed automatically. The competitive immunoassay for clenbuterol was performed. The free clenbuterol in the sample competed with HRP-clenbuterol to bind with the anti-clenbuterol antibody coated on analysis channel. After adding the enzyme substrate, the fluorescence signal was detected by a linear confocal laser-induced fluorescence scanner. The linear range and LOD of clenbuterol were 0–5.0 ng/mL and 0.088 ng/mL, respectively.

4. Conclusions and Perspectives

In the past few decades, the illegal use of β_2-agonists in livestock has received continuous attention worldwide due to its potential threat to public health. In this review, we summarized antibody-based techniques to extract β_2-agonists from samples, and focused on the development and application of immunological methods in detecting β_2-agonists (Table 2). Most of the methods can measure β_2-agonists with high specificity and sensitivity. RIA is replaced largely by nonisotopic immunoassays, due to the difficulties associated with the handling and storage of radioactive reagents and disposal of radioactive waste. ELISA can be used for high-throughput detection of β_2-agonists, but the sensitivity is moderate and the assay requires a relatively long time. Introduction of nanomaterials in ELISA as enzyme carriers, enzyme mimics and signal reporters may improve the sensitivity, stability and measurement time of the assay. Combination of ELISA with other methods, such as SERS, can also improve the sensitivity of detection. LFIA is fast, easy to perform, not dependent on special equipment and can be carried out on-site to qualitatively or semiquantitatively detect β_2-agonists. The use of fluorescent nanomaterials, quantum dots and the upconverting phosphors as probes in LFIA can improve the sensitivity of detection. SERS-based LFIA further increase the sensitivity for detecting β_2-agonists, but instrumentation is required. Chemiluminescence immunoassay (CLIA) and immunosensors, such as SPR-, SERS-, ECL- and electrochemistry-based immunosensors, can detect β_2-agonists in a short time with high sensitivity, but SPR- and SERS-based immunoassays require special instruments. Current CLIAs for detecting the β_2-agonist use enzymes to label the secondary Ab and luminescent chemical as substrate to generate light emission. The introduction of luminescent nanoparticles as hemiluminescence probe, such as Au nanoparticles, QDs and magnetic materials, may enhance the sensitivity of the assay. The conventional SPR immunosensors for detecting β_2-agonists are simple and rapid, but the sensitivity is not adequate. The introduction of Au nanomaterials as labels for Ab and as the amplification tags immobilized on the sensor surface enhances the SPR signal. Continued research should be pursued to explore novel nanomaterials to improve the performance, to generate the SPR chips for recyclable use and exploit miniaturized SPR devices to reduce cost. SERS immunosensors based on nanoprobes detect β_2-agonists with high sensitivity. However, their stability and reproducibility need to be further improved. Current ECL immunosensors for β_2-agonist detection mainly use QDs as probes. Synthesis of new ECL probes, development of new biointerface construction and new signal amplification strategies will improve the performance of sensors. In electrochemical biosensors to detect β_2-agonists, nanomaterials, such as gold nanoparticles and graphene, have been used to modify the electrodes or label antibodies to increase conductivity, electron transfer and signal generation. The performance of the immunosensors may be improved by the construction of labels with a high loading of signal species, introduction of interfacial reaction initiated by functionalized nanomaterials and building a synergistic connection between labels and substrate. More research is needed to develop stable, selective, sensitive, rapid and portable devices for detecting β_2-agonists.

Table 2. Immunoassays for the detection of β_2-agonists.

Analytical Technologies	Samples	Analyte	Limit of Detection	References
1. Radioimmunoassay	human plasma	salbutamol	0.5 ng/mL	[30]
	plasma, urine	fenoterol	10–20 pg/mL	[31]
	horse urine	albuterol	28.8 fmol/tube	[32]
	cattle plasma, urine, feces	clenbuterol	7.8 pg/tube	[33]
	bovine liver	clenbuterol, mabuterol, etc.	0.1 μg/kg	[34]
		brombuterol, cimbuterol, etc.	0.3 μg/kg	
2. ELISA				
2.1. Direct competitive ELISA	pork	clenbuterol	0.09 ng/g	[35]
	milk	clenbuterol	0.045 ng/mL	[36]
	feed, milk, swine urine,	clenbuterol	0.03 ng/mL	[39]
	swine serum	salbutamol	0.25 ng/mL	[37]
	swine urine	phenylethanolamine A	0.5 μg/L	[38]
2.2 Indirect competitive ELISA	swine urine, pork	phenylethanolamine A	urine: 0.13 ng/mL, pork: 0.39 ng/g	[40]
	sheep urine	ractopamine	0.35 ng/mL	[41]
	swine and bovine urine	zilpaterol	IC_{50} 3.94 ± 0.48 ng/mL	[42]
	clenbuterol solution	clenbuterol	0.50 ng/mL	[43]
	clenbuterol solution	clenbuterol	0.3 pg/mL	[44]
	salbutamol solution, urine	salbutamol	0.04 ng/mL	[45]
	livestock wastewater	salbutamol	0.66 ng/L	[46]
3. Chemiluminescence immunoassay	pork, liver	salbutamol	0.15 ng/mL	[47]
	swine meat, feed	brombuterol	0.33 pg/mL	[48]
	swine urine	ractopamine	0.97 ng/mL	[49]
4. Lateral flow immunoassay (LFIA)				
4.1. Colorimetric LFIA				
4.1.1. Gold nanoparticle as tracer	swine urine	clenbuterol, ractopamine	0.1 ± 0.01 ng/mL	[50]
	pork muscle	clenbuterol	0.10 ng/g	[51]
	swine urine	salbutamol	1.0 ng/mL	[52]
	swine and turkey meat, cow milk	salbutamol	meat: 3.0 ng/g; milk: 4.0 ng/g	[53]
	swine urine, pork	phenylethanolamine A	5 ng/mL(g)	[40]
	urine, pork	phenylethanolamine A	0.1 ng/mL	[54]
	swine urine	ractopamine	0.13 ng/mL	[55]
	swine urine	ractopamine	2 ng/mL	[56]
	turkey meat, beef liver	ractopamine	0.5 ng/mL	[57]

Table 2. *Cont.*

Analytical Technologies	Samples	Analyte	Limit of Detection	References
4.1.1. Gold nanoparticle as tracer	swine feed	ractopamine	0.1 ng/g	[58]
	swine urine and feed, milk	clenbuterol	urine: 0.1 ng/mL; feed: 0.2 ng/g; milk: 0.5 ng/mL	[59]
	clenbuterol solution	clenbuterol	2 ppb	[60]
	pork	clenbuterol	0.04 ng/mL	[61]
4.1.2. Other nanoparticles as tracers	swine urine, pork	clenbuterol	urine: 6 ng/mL; pork: 5 ng/mL	[62]
	clenbuterol, ractopamine solution	clenbuterol, ractopamine	clenbuterol: 3 ng/mL; ractopamine: 2 ng/mL	[63]
	swine urine	clenbuterol	3 ng/mL	[64]
	swine urine	ractopamine, salbutamol	ractopamine: 1.0 ng/mL; salbutamol: 3.0 ng/mL	[65]
	pork, swine kidney and bacon	clenbuterol	pork: 3 ng/g; kidney and bacon: 5 ng/g	[66]
	swine feed, pork	ractopamine	feed: 2.0 ng/mL; pork: 1.0 ng/mL	[67]
4.2. Luminescent LFIA	swine urine	clenbuterol	0.037 ng/mL	[68]
	swine urine	clenbuterol	0.22 ng/mL	[29]
	swine urine, feed, pork	clenbuterol, ractopamine, salbuterol	clenbuterol: 0.10 ng/mL; ractopamine: 0.10 ng/mL; salbuterol: 0.09 ng/mL	[69]
	swine urine	ractopamine	7.2 pg/mL	[71]
	swine urine, muscle	ractopamine	0.16 ng/mL	[72]
	pork tissue, urine, feed	clenbuterol	0.01 ng/mL	[73]
	swine urine	ractopamine, clenbuterol	ractopamine: 0.17 ng/mL; clenbuterol: 0.067 ng/mL	[74]
4.3. Other types of LFIA	milk, swine liver, tenderloin	clenbuterol	2 ng/mL	[75]
	swine urine	phenylethanolamine A	0.32 pg/mL	[76]
	swine urine	clenbuterol	0.24 pg/mL	[77]
	swine meat, urine	brombuterol	0.5 pg/mL	[78]
5. Immunosensors				
5.1. Surface plasmon resonance sensors	salbutamol solution	salbutamol	5 ng/mL	[80]
	ractopamine and salbutamol solution	ractopamine, salbutamol	ractopamine: 10 pg/mL, salbutamol: 5 pg/mL	[81]
	swine urine	ractopamine	0.09 ng/mL	[82]
	clenbuterol solution	clenbuterol	0.05 pg/mL	[83]
	bovine urine	clenbuterol	100 fg/mL	[84]

Table 2. *Cont.*

Analytical Technologies	Samples	Analyte	Limit of Detection	References
5.2. SERS-based immunosensor	swine urine	clenbuterol	0.1 pg/mL	[85]
	clenbuterol and ractopamine solution	clenbuterol, ractopamine	1.0 pg/mL	[86]
	swine meat and liver, human urine	salbutamol, brombuterol	salbutamol 2.0 pg/mL; brombuterol 1.0.pg/mL	[87]
	clenbuterol solution	clenbuterol	0.22 fg/mL	[88]
	clenbuterol solution	clenbuterol	0.68 pg/mL	[89]
5.3. Electrochemiluminescence immunosensor	pork and liver	salbutamol	8.4 pg/mL	[91]
	pork and liver	salbutamol	17 pg/mL	[92]
	pork and feed	ractopamine	1.7 pg/mL	[93]
	pork	salbutamol	0.17 pg/mL	[94]
	pork and feed	brombuterol	1.5 pg/mL	[95]
	pork	brombuterol	0.3 pg/mL	[96]
	pork	ractopamine	2.6 pg/mL	[97]
	pork and swine feed extract	brombuterol	0.31 pg/mL	[98]
5.4. Electrochemical immunosensor	porcine serum	salbutamol	0.2 fg/mL	[99]
	serum	salbutamol	7 pg/mL	[102]
	swine urine	ractopamine	2.3 pg/mL	[100]
	clenbuterol solution	clenbuterol	0.12 ng/mL	[101]
	pork	salbutamol, ractopamine, clenbuterol	salbutamol: 1.44 pg/mL, clenbuterol: 1.38 pg/mL, ractopamine: 1.52 pg/mL	[103]
	pork, feed	salbutamol	0.04 ng/mL	[105]
	salbutamol solution	salbutamol	0.06 ng/mL	[106]
	salbutamol solution	salbutamol	0.03 ng/mL	[107]
	bovine hair	clenbuterol	0.008 ng/mL	[108]
	milk	clenbuterol	0.196 ng/mL	[109]
	clenbuterol solution	clenbuterol	0.076 ng/mL	[111]
	swine feed	clenbuterol	0.25 ng/mL	[110]
6. Other types of immunoassays	human urine	salbutamol	28 fg/mL	[113]
	water, urine	salbutamol	0.65 fmol/ML	[114]
	pork	salbutamol, ractopamine	sabutamol: 0.051 µg/kg, ractopamine: 0.02 µg/kg	[115]
	turkey meat	ractopamine	1 ng/mL	[116]
	pork	ractopamine	0.56 µg/kg	[117]
	clenbuterol, ractopamine and salbutamol solution	clenbuterol, ractopamine, salbutamol	clenbuterol: 0.09 µg/L; ractopamine: 0.50 µg/L; salbutamol: 0.01 µg/L	[118]
	swine urine	clenbuterol	0.088 ng/mL	[119]

Author Contributions: Conceptualization, S.O., S.Y. and Y.L.; writing—original draft preparation, S.O. and S.Y.; writing—review and editing, Y.L.; funding acquisition, Y.L. All authors have read and agreed to the published version of the manuscript.

Funding: This work was supported by the National Key Research and Development Program of China (grant number 2017YFC1601702).

Institutional Review Board Statement: Not applicable.

Informed Consent Statement: Not applicable.

Data Availability Statement: Not applicable.

Conflicts of Interest: The authors declare no conflict of interest.

References

1. Billington, C.K.; Penn, R.B.; Hall, I.P. β2 agonists. *Handb. Exp. Pharmacol.* **2017**, *237*, 23–40. [PubMed]
2. Sillence, M.N. Technologies for the control of fat and lean deposition in livestock. *Vet. J.* **2004**, *167*, 242–257. [CrossRef] [PubMed]
3. Mazzanti, G.; Daniele, C.; Boatto, G.; Manca, G.; Brambilla, G.; Loizzo, A. New beta-adrenergic agonists used illicitly as growth promoters in animal breeding: Chemical and pharmacodynamic studies. *Toxicology* **2003**, *187*, 91–99. [CrossRef]
4. Prezelj, A.; Obreza, A.; Pecar, S. Abuse of clenbuterol and its detection. *Curr. Med. Chem.* **2003**, *10*, 281–290. [CrossRef] [PubMed]
5. Ramos, F.; Baeta, M.L.; Reis, J.; Silveira, M.I. Evaluation of the illegal use of clenbuterol in Portuguese cattle farms from drinking water, urine, hair and feed samples. *Food Addit. Contam. Part A Chem. Anal. Control Expo. Risk Assess.* **2009**, *26*, 814–820. [CrossRef]
6. Sakai, N.; Sakai, M.; Mohamad Haron, D.E.; Yoneda, M.; Ali Mohd, M. Beta-agonist residues in cattle, chicken and swine livers at the wet market and the environmental impacts of waste water from livestock farms in Selangor State, Malaysia. *Chemosphere* **2016**, *165*, 183–190. [CrossRef] [PubMed]
7. Martínez-Navarro, J.F. Food poisoning related to consumption of illicit beta-agonist in liver. *Lancet* **1990**, *336*, 1311. [CrossRef]
8. Brambilla, G.; Loizzo, A.; Fontana, L.; Strozzi, M.; Guarino, A.; Soprano, V. Food poisoning following consumption of clenbuterol-treated veal in Italy. *J. Am. Med. Assoc.* **1997**, *278*, 635. [CrossRef]
9. Sporano, V.; Grasso, L.; Esposito, M.; Oliviero, G.; Brambilla, G.; Loizzo, A. Clenbuterol residues in non-liver containing meat as a cause of collective food poisoning. *Vet. Hum. Toxicol.* **1998**, *40*, 141–143. [PubMed]
10. Brambilla, G.; Cenci, T.; Franconi, F.; Galarini, R.; Macrì, A.; Rondoni, F.; Strozzi, M.; Loizzo, A. Clinical and pharmacological profile in a clenbuterol epidemic poisoning of contaminated beef meat in Italy. *Toxicol. Lett.* **2000**, *114*, 47–53. [CrossRef]
11. Wu, M.L.; Deng, J.F.; Chen, Y.; Chu, W.L.; Hung, D.Z.; Yang, C.C. Late diagnosis of an outbreak of leanness-enhancing agent-related food poisoning. *J. Am. J. Emerg. Med.* **2013**, *31*, 1501–1503. [CrossRef] [PubMed]
12. Baynes, R.E.; Dedonder, K.; Kissell, L.; Mzyk, D.; Marmulak, T.; Smith, G.; Tell, L.; Gehring, R.; Davis, J.; Riviere, J.E. Health concerns and management of select veterinary drug residues. *Food Chem. Toxicol.* **2016**, *88*, 112–122. [CrossRef]
13. Boyd, D.; O'Keeffe, M.; Smyth, M.R. Methods for the determination of beta-agonists in biological matrices. *Rev. Anal.* **1996**, *121*, 1R–10R. [CrossRef] [PubMed]
14. Zhang, W.; Wang, P.; Su, X. Current advancement in analysis of β-agonists. *TrAC Trends Anal. Chem.* **2016**, *85*, 1–16. [CrossRef]
15. Li, G.; Zhang, X.; Zheng, F.; Liu, J.; Wu, D. Emerging nanosensing technologies for the detection of β-agonists. *Food Chem.* **2020**, *332*, 127431. [CrossRef] [PubMed]
16. Danyi, S.; Degand, G.; Duez, C.; Granier, B.; Maghuin-Rogister, G.; Scippo, M.L. Solubilisation and binding characteristics of a recombinant beta2-adrenergic receptor expressed in the membrane of Escherichia coli for the multianalyte detection of beta-agonists and antagonists residues in food-producing animals. *Anal. Chim. Acta* **2007**, *589*, 159–165. [CrossRef]
17. Cheng, G.; Li, F.; Peng, D.; Huang, L.; Hao, H.; Liu, Z.; Wang, Y.; Yuan, Z. Development of an enzyme-linked-receptor assay based on Syrian hamster β2-adrenergic receptor for detection of β-agonists. *Anal. Biochem.* **2014**, *459*, 18–23. [CrossRef] [PubMed]
18. dos Ramos, F.J. Beta2-agonist extraction procedures for chromatographic analysis. *J. Chromatogr. A* **2000**, *880*, 69–83. [CrossRef]
19. Haasnoot, W.; Ploum, M.E.; Paulussen, R.J.; Schilt, R.; Huf, F.A. Rapid determination of clenbuterol residues in urine by high-performance liquid chromatography with on-line automated sample processing using immunoaffinity chromatography. *J. Chromatogr.* **1990**, *519*, 323–335. [CrossRef]
20. Cai, J.; Henion, J. Quantitative multi-residue determination of beta-agonists in bovine urine using on-line immunoaffinity extraction-coupled column packed capillary liquid chromatography-tandem mass spectrometry. *J. Chromatogr. B Biomed. Sci. Appl.* **1997**, *691*, 357–370. [CrossRef]
21. Shelver, W.L.; Smith, D.J. Immunoaffinity column as sample cleanup method for determination of the beta-adrenergic agonist ractopamine and its metabolites. *J. AOAC Int.* **2002**, *85*, 1302–1307. [CrossRef] [PubMed]
22. Lin, Q.B.; Zhao, X.T.; Song, H.; Pan, Y.L. Immunoaffinity chromatography purification and ultra-high-performance liquid chromatography-tandem mass spectrometry determination of four β-agonists in beef. *Food Addit. Contam. Part A Chem. Anal. Control Expo. Risk Assess.* **2012**, *29*, 935–941. [CrossRef] [PubMed]

23. Hellenäs, K.-E.; Johansson, M.A.; Elliott, C.T.; Sternesjö, A.; Stenberg, E. *EuroResidue IV-Proceedings of the Conference on Residues of Veterinary Drugs in Food*; van Ginkel, L.A., Ruiter, A., Eds.; RIVM, Laboratory for Residue Analysis: Bilthoven, The Netherlands, 2000; pp. 542–545.
24. Haasnoot, W.; Kemmers-Voncken, A.; van Rhijn, H.; Schilt, R. Immunofiltration as al ternative for immunoaffinity chromatography. In *EuroResidue IV-Proceedings of the Conference on Residues of Veterinary Drugs in Food*; van Ginkel, L.A., Ruiter, A., Eds.; RIVM, Laboratory for Residue Analysis: Bilthoven, The Netherlands, 2000; pp. 496–500.
25. Haasnoot, W.; Kemmers-Voncken, A.; Samson, D. Immunofiltration as sample cleanup for the immunochemical detection of beta-agonists in urine. *Analyst* **2002**, *127*, 87–92. [CrossRef]
26. Cheng, J.; Su, X.O.; Wang, S.; Zhao, Y. Highly Sensitive detection of clenbuterol in animal urine using immunomagnetic bead treatment and surface-enhanced Raman spectroscopy. *Sci. Rep.* **2016**, *6*, 32637. [CrossRef] [PubMed]
27. Cheng, J.; Su, X.O.; Han, C.; Wang, S.; Wang, P.; Zhang, S.; Xie, J. Ultrasensitive detection of salbutamol in animal urine by immunomagnetic bead treatment coupling with surface-enhanced Raman spectroscopy. *Sens. Actuators B Chem.* **2018**, *255*, 2329–2338. [CrossRef]
28. Peng, T.; Wang, J.; Zhao, S.; Xie, S.; Yao, K.; Zheng, P.; Wang, S.; Ke, Y.; Jiang, H. A fluorometric clenbuterol immunoassay based on the use of organic/inorganic hybrid nanoflowers modified with gold nanoclusters and artificial antigen. *Mikrochim. Acta* **2018**, *185*, 366. [CrossRef]
29. Huang, Z.; Xiong, Z.; Chen, Y.; Hu, S.; Lai, W. Sensitive and matrix-tolerant lateral flow immunoassay based on fluorescent magnetic nanobeads for the detection of clenbuterol in swine urine. *J. Agric. Food Chem.* **2019**, *67*, 3028–3036. [CrossRef] [PubMed]
30. Loo, J.C.; Beaulieu, N.; Jordan, N.; Brien, R.; McGilveray, I.J. A specific radio-immunoassay (RIA) for salbutamol (albuterol) in human plasma. *Res. Commun. Chem. Pathol. Pharmacol.* **1987**, *55*, 283–286. [PubMed]
31. Rominger, K.L.; Mentrup, A.; Stiasni, M. Radioimmunological determination of fenoterol. Part II: Antiserum and tracer for the determination of fenoterol. *Arzneimittelforsch* **1990**, *40*, 887–895. [PubMed]
32. Adam, A.; Ong, H.; Sondag, D.; Rapaille, A.; Marleau, S.; Bellemare, M.; Raymond, P.; Giroux, D.; Loo, J.K.; Beaulieu, N. Radioimmunoassay for albuterol using a monoclonal antibody: Application for direct quantification in horse urine. *J. Immunoass.* **1990**, *11*, 329–345. [CrossRef] [PubMed]
33. Delahaut, P.; Dubois, M.; Pri-Bar, I.; Buchman, O.; Degand, G.; Ectors, F. Development of a specific radioimmunoassay for the detection of clenbuterol residues in treated cattle. *Food Addit. Contam.* **1991**, *8*, 43–53. [CrossRef] [PubMed]
34. Granja, R.H.; Montes, N.A.M.; Rabone, F.; Montes Niño, R.E.; Cannavan, A.; Salerno, A.G. Validation of radioimmunoassay screening methods for beta-agonists in bovine liver according to Commission Decision 2002/657/EC. *Food Addit. Contam. Part A Chem. Anal. Control Expo. Risk Assess.* **2008**, *25*, 1475–1481. [CrossRef]
35. Bui, Q.A.; Vu, T.H.; Ngo, V.K.; Kennedy, I.R.; Lee, N.A.; Allan, R. Development of an ELISA to detect clenbuterol in swine products using a new approach for hapten design. *Anal. Bioanal. Chem.* **2016**, *408*, 6045–6052. [CrossRef] [PubMed]
36. Talib, N.A.A.; Salam, F.; Sulaiman, Y. Development of polyclonal antibody against clenbuterol for immunoassay application. *Molecules* **2018**, *23*, 789. [CrossRef] [PubMed]
37. Lei, Y.C.; Tsai, Y.F.; Tai, Y.T.; Lin, C.Y.; Hsieh, K.H.; Chang, T.H.; Sheu, S.Y.; Kuo, T.F. Development and fast screening of salbutamol residues in swine serum by an enzyme-linked immunosorbent assay in Taiwan. *J. Agric. Food Chem.* **2008**, *56*, 5494–5499. [CrossRef] [PubMed]
38. Bai, Y.; Liu, Z.; Bi, Y.; Wang, X.; Jin, Y.; Sun, L.; Wang, H.; Zhang, C.; Xu, S. Preparation of polyclonal antibodies and development of a direct competitive enzyme-linked immunosorbent assay to detect residues of phenylethanolamine A in urine samples. *J. Agric. Food Chem.* **2012**, *60*, 11618–11624. [CrossRef] [PubMed]
39. Huang, Q.; Zhao, B.; Bold, S.E.; Bu, T.; Yan, L.; Dou, L.; Yang, Q.; Wang, J.; Zhang, D. Immunoassay of clenbuterol with bacteria as natural signal carriers for signal amplification. *Sens. Actuators B Chem.* **2019**, *288*, 210–216. [CrossRef]
40. Li, X.; Wang, W.; Wang, L.; Wang, Q.; Pei, X.; Jiang, H. Rapid determination of phenylethanolamine A in biological samples by enzyme-linked immunosorbent assay and lateral-flow immunoassay. *Anal. Bioanal. Chem.* **2015**, *407*, 7615–7624. [CrossRef] [PubMed]
41. Han, S.; Zhou, T.; Yin, B.; He, P. Gold nanoparticle-based colorimetric ELISA for quantification of ractopamine. *Mikrochim. Acta* **2018**, *185*, 210. [CrossRef]
42. Shelver, W.L.; Smith, D.J. Enzyme-linked immunosorbent assay development for the beta-adrenergic agonist zilpaterol. *J. Agric. Food Chem.* **2004**, *52*, 2159–2166. [CrossRef]
43. Lu, Q.; Hou, Y.Y.; Liu, X.X.; Wang, H.; Hou, J.J.; Wei, J.L.; Zhou, S.S.; Liu, X.Y. Construction, expression and functional analysis of anti-clenbuterol codon-optimized scFv recombinant antibody. *Food Chem. Toxicol.* **2020**, *135*, 110973. [CrossRef] [PubMed]
44. Wang, L.; Wang, J.; Zhang, A.; Huang, X.A.; Lei, H. Two binding epitopes modulating specificity of immunoassay for β-agonist detection: Quantitative structure-activity relationship. *Food Chem.* **2022**, *371*, 131071. [CrossRef] [PubMed]
45. Wang, L.; Jiang, W.; Shen, X.; Li, X.; Huang, X.A.; Xu, Z.; Sun, Y.; Chan, S.W.; Zeng, L.; Eremin, S.A.; et al. Four hapten spacer sites modulating class specificity: Nondirectional multianalyte immunoassay for 31 β-agonists and analogues. *Anal. Chem.* **2018**, *90*, 2716–2724. [CrossRef] [PubMed]
46. Fang, S.; Zhang, Y.; Liu, X.; Qiu, J.; Liu, Z.; Kong, F. Development of a highly sensitive time-resolved fluoroimmunoassay for the determination of trace salbutamol in environmental samples. *Sci. Total Environ.* **2019**, *679*, 359–364. [CrossRef] [PubMed]

47. Xu, M.; Qian, X.; Zhao, K.; Deng, A.; Li, J. Flow injection chemiluminescent competitive immunoassay for the β-adrenergic agonist salbutamol using carboxylic resin beads and enzymatic amplification. *Sens. Actuator B Chem.* **2015**, *215*, 323–329. [CrossRef]
48. Zhou, X.; Li, Y.; Shi, J.; Zhao, K.; Deng, A.; Li, J. A flow injection chemiluminescent immunosensor for ultrasensitive detection of brombuterol based on resin beads and enzymatic amplification. *Food Anal. Methods* **2019**, *12*, 305–312. [CrossRef]
49. Wang, S.; Chen, Q.; Wei, X.; Wu, J.; Wang, C.; Liu, J.; Zhang, L.; Dong, Y. A competitive luminol chemiluminescence immunosensor based on a microfluidic chip for the determination of ractopamine. *Electrophoresis* **2017**, *38*, 368–371. [CrossRef]
50. Zhang, M.Z.; Wang, M.Z.; Chen, Z.L.; Fang, J.H.; Fang, M.M.; Liu, J.; Yu, X.P. Development of a colloidal gold-based lateral-flow immunoassay for the rapid simultaneous detection of clenbuterol and ractopamine in swine urine. *Anal. Bioanal. Chem.* **2009**, *395*, 2591–2599. [CrossRef]
51. Wu, K.; Guo, L.; Xu, W.; Xu, H.; Aguilar, Z.P.; Xu, G.; Lai, W.; Xiong, Y.; Wan, Y. Sulfonated polystyrene magnetic nanobeads coupled with immunochromatographic strip for clenbuterol determination in pork muscle. *Talanta* **2014**, *129*, 431–437. [CrossRef]
52. Xie, C.H.; Chen, F.J.; Yang, T.B. A high-affinity anti-salbutamol monoclonal antibody: Key to a robust lateral-flow immunochromatographic assay. *Anal. Biochem.* **2012**, *426*, 118–125. [CrossRef]
53. Zvereva, E.A.; Zherdev, A.V.; Xu, C.; Dzantiev, B.B. Highly sensitive immunochromatographic assay for qualitative and quantitative control of beta-agonist salbutamol and its structural analogs in foods. *Food Control* **2018**, *86*, 50–58. [CrossRef]
54. Li, J.H.; Li, C.S.; Wu, M.; Zhang, Y.; Ma, X.F.; Cheng, H.; Yan, J.H. Development of an ultrasensitive immunochromatographic assay (ICA) strip for the rapid detection of phenylethanolamine A in urine and pork samples. *J. Food Sci.* **2015**, *80*, 894–899.
55. Ren, M.L.; Chen, X.L.; Li, C.H.; Xu, B.; Liu, W.J.; Xu, H.Y.; Xiong, Y.H. Lateral flow immunoassay for quantitative detection of ractopamine in swine urine. *Biomed. Environ. Sci.* **2014**, *27*, 134–137. [PubMed]
56. Gu, H.Y.; Liu, L.P.; Song, S.S.; Kuang, H.; Xu, C.L. Development of an immunochromatographic strip assay for ractopamine detection using an ultrasensitive monoclonal antibody. *Food Agric. Immunol.* **2016**, *27*, 471–483. [CrossRef]
57. Zvereva, E.A.; Shpakova, N.A.; Zherdev, A.V.; Xu, C.; Dzantiev, B.B. Highly sensitive immunochromatographic assay for qualitative and quantitative control of beta-agonist ractopamine in Foods. *Appl. Biochem. Microbiol.* **2018**, *54*, 436–441. [CrossRef]
58. Preechakasedkit, P.; Ngamrojanavanich, N.; Khongchareonporn, N.; Chailapakul, O. Novel ractopamine-protein carrier conjugation and its application to the lateral flow strip test for ractopamine detection in animal feed. *J. Zhejiang Univ. Sci. B* **2019**, *20*, 193–204. [CrossRef]
59. Huang, Q.; Bu, T.; Zhang, W.; Yan, L.; Zhang, M.; Yang, Q.; Huang, L.; Yang, B.; Hu, N.; Suo, Y.; et al. An improved clenbuterol detection by immunochromatographic assay with bacteria@Au composite as signal amplifier. *Food Chem.* **2018**, *262*, 48–55. [CrossRef]
60. Wang, J.; Zhang, L.; Huang, Y.; Dandapat, A.; Dai, L.; Zhang, G.; Lu, X.; Zhang, J.; Lai, W.; Chen, T. Hollow Au-Ag nanoparticles labeled immunochromatography strip for highly snsitive detection of clenbuterol. *Sci. Rep.* **2017**, *7*, 41419. [CrossRef]
61. Chen, Y.; Huang, Z.; Hu, S.; Zhang, G.; Peng, J.; Xia, J.; Lai, W. Integrated immunochromatographic assay for qualitative and quantitative detection of clenbuterol. *Anal. Biochem.* **2019**, *577*, 45–51. [CrossRef]
62. Zhu, C.J.; Zhao, G.Y.; Dou, W.C. Immunochromatographic assay using brightly colored silica nanoparticles as visible label for point-of-care detection of clenbuterol. *Sens. Actuators B* **2018**, *266*, 392–399. [CrossRef]
63. Yu, Q.; Liu, J.; Zhao, G.; Dou, W. A silica nanoparticle based 2-colorimmunochromatographic assay for simultaneous determination of clenbuterol and ractopamine. *Mikrochim. Acta* **2019**, *186*, 421. [CrossRef] [PubMed]
64. Wang, Z.Z.; Jing, J.; Ren, Y.G.; Guo, Y.F.; Tao, N.Y.; Zhou, Q.W.; Zhang, H.L.; Ma, Y.F.; Wang, Y.H. Preparation and application of selenium nanoparticles in a lateral flow immunoassay for clenbuterol detection. *Mater. Lett.* **2019**, *234*, 212–215. [CrossRef]
65. Wang, Z.; Zhou, Q.; Guo, Y.; Hu, H.; Zheng, Z.; Li, S.; Wang, Y.; Ma, Y. Rapid detection of ractopamine and salbutamol in swine urine by immunochromatography based on selenium nanoparticles. *Int. J. Nanomed.* **2021**, *16*, 2059–2070. [CrossRef] [PubMed]
66. Zhao, B.X.; Huang, Q.; Dou, L.N.; Bu, T.; Chen, K.; Yang, Q.F.; Yan, L.Z.; Wang, J.L.; Zhang, D.H. Prussian blue nanoparticles based lateral flow assay for high sensitive determination of clenbuterol. *Sens. Actuators B* **2018**, *275*, 223–229. [CrossRef]
67. Liu, J.; Yu, Q.; Zhao, G.; Dou, W. Ultramarine blue nanoparticles as a label for immunochromatographic on-site determination of ractopamine. *Mikrochim. Acta* **2020**, *187*, 285. [CrossRef]
68. Song, C.; Zhi, A.; Liu, Q.; Yang, J.; Jia, G.; Shervin, J.; Tang, L.; Hu, X.; Deng, R.; Xu, C.; et al. Rapid and sensitive detection of β-agonists using a portable fluorescence biosensor based on fluorescent nanosilica and a lateral flow test strip. *Biosens. Bioelectron.* **2013**, *50*, 62–65. [CrossRef]
69. Wang, P.; Wang, Z.; Su, X. A sensitive and quantitative fluorescent multi-component immuno-chromatographic sensor for β-agonist residues. *Biosens. Bioelectron.* **2015**, *64*, 511–516. [CrossRef]
70. Wang, R.; Zhang, W.; Wang, P.; Su, X. A paper-based competitive lateral flow immunoassay for multi β-agonist residues by using a single monoclonal antibody labelled with red fluorescent nanoparticles. *Mikrochim. Acta* **2018**, *185*, 191. [CrossRef]
71. Hu, L.M.; Luo, K.; Xia, J.; Xu, G.M.; Wu, C.H.; Han, J.J.; Zhang, G.G.; Liu, M.; Lai, W.H. Advantages of time-resolved fluorescent nanobeads compared with fluorescent submicrospheres, quantum dots, and colloidal gold as label in lateral flow assays for detection of ractopamine. *Biosens. Bioelectron.* **2017**, *91*, 95–103. [CrossRef]
72. Shi, C.Y.; Deng, N.; Liang, J.J.; Zhou, K.N.; Fu, Q.Q.; Tang, Y. A fluorescent polymer dots positive readout fluorescent quenching lateral flow sensor for ractopamine rapid detection. *Anal. Chim. Acta* **2015**, *854*, 202–208. [CrossRef]
73. Wang, P.; Wang, R.; Zhang, W.; Su, X.; Luo, H. Novel fabrication of immunochromatographic assay based on up conversion phosphors for sensitive detection of clenbuterol. *Biosens. Bioelectron.* **2016**, *77*, 866–870. [CrossRef] [PubMed]

74. Wang, W.; Su, X.; Ouyang, H.; Wang, L.; Fu, Z. A novel immunochromatographic assay based on a time-resolved chemiluminescence strategy for the multiplexed detection of ractopamine and clenbuterol. *Anal. Chim. Acta* **2016**, *917*, 79–84. [CrossRef] [PubMed]
75. Zhang, H.; Wang, L.; Yao, X.; Wang, Z.; Dou, L.; Su, L.; Zhao, M.; Sun, J.; Zhang, D.; Wang, J. Developing a simple immunochromatography assay for clenbuterol with sensitivity by one-step staining. *J. Agric. Food Chem.* **2020**, *68*, 15509–15515. [CrossRef] [PubMed]
76. Li, M.; Yang, H.; Li, S.; Zhao, K.; Li, J.; Jiang, D.; Sun, L.; Deng, A. Ultrasensitive and quantitative detection of a new β-agonist phenylethanolamine A by a novel immunochromatographic assay based on surface-enhanced Raman scattering (SERS). *J. Agric. Food Chem.* **2014**, *62*, 10896–10902. [CrossRef] [PubMed]
77. Xie, Y.; Chang, H.F.; Zhao, K.; Li, J.G.; Yang, H.; Mei, L.Y.; Xu, S.M.; Deng, A.P. A novel immunochromatographic assay (ICA) based on surface-enhanced Raman scattering for the sensitive and quantitative determination of clenbuterol. *Anal. Methods* **2015**, *7*, 513–520. [CrossRef]
78. Fu, H.Q.; Chu, Y.X.; Zhao, K.; Li, J.G.; Deng, A.P. Ultrasensitive detection of the β-adrenergic agonist brombuterol by a SERS-based lateral flow immunochromatographic assay using flower-like gold-silver core-shell nanoparticles. *Microchim. Acta* **2017**, *184*, 1711–1719. [CrossRef]
79. Nguyen, H.H.; Park, J.; Kang, S.; Kim, M. Surface plasmon resonance: A versatile technique for biosensor applications. *Sensors* **2015**, *15*, 10481–10510. [CrossRef]
80. Li, W.; Wang, C.; Li, H.; Chen, Z.; Yan, M. Strategy of Dimercaptothiol as self-assembled monolayers enhance the sensitivity of SPR immunosensor for detection of salbutamol. *Anal. Sci.* **2021**, *37*, 1289–1294. [CrossRef]
81. Suherman, M.K.; Kawaguchi, T. Highly selective and sensitive detection of β-agonists using a surface plasmon resonance sensor based on an alkanethiol monolayer functionalized on a Au surface. *Biosens. Bioelectron.* **2015**, *67*, 356–363. [CrossRef]
82. Wang, S.; Zhao, S.; Wei, X.; Zhang, S.; Liu, J.; Dong, Y. An improved label-free indirect competitive SPR immunosensor and its comparison with conventional ELISA for ractopamine detection in swine urine. *Sensors* **2017**, *17*, 604. [CrossRef]
83. Kabiraz, D.C.; Morita, K.; Sakamoto, K.; Kawaguchi, T. Mechanism of surface plasmon resonance sensing by indirect competitive inhibition immunoassay using Au nanoparticle labeled antibody. *Talanta* **2017**, *172*, 1–7. [CrossRef] [PubMed]
84. Kabiraz, D.C.; Morita, K.; Sakamoto, K.; Takahashi, M.; Kawaguchi, T. Highly sensitive detection of clenbuterol in urine sample by using surface plasmon resonance immunosensor. *Talanta* **2018**, *186*, 521–526. [CrossRef] [PubMed]
85. Zhu, G.C.; Hu, Y.J.; Gao, J.; Zhong, L. Highly sensitive detection of clenbuterol using competitive surface-enhanced Raman scattering immunoassay. *Anal. Chim. Acta* **2011**, *697*, 61–66. [CrossRef]
86. Yu, M.; Hu, Y.J.; Liu, J.Z. Simultaneous detection of clenbuterol and ractopamine based on multiplexed competitive surface enhanced Raman scattering (SERS) immunoassay. *New J. Chem.* **2017**, *41*, 10407. [CrossRef]
87. Gu, X.F.; Tian, S.; Chen, Y.Y.; Wang, Y.X.; Gu, D.D.; Guo, E.H.; Liu, Y.; Li, J.G.; Deng, A.P. A SERS-based competitive immunoassay using highly ordered gold cavity arrays as the substrate for simultaneous detection of β-adrenergic agonists. *Sens. Actuators B Chem.* **2021**, *345*, 130230. [CrossRef]
88. Wei, C.; Zhang, C.J.; Xu, M.M.; Yuan, Y.X.; Yao, J.L. Liquid magnetic competitive immunoassay ofclenbuterol based on surface-enhancedRaman spectroscopy. *J. Raman Spectrosc.* **2017**, *48*, 1307–1317. [CrossRef]
89. Yao, D.; Li, C.; Wen, G.; Liang, A.; Jiang, Z. A highly sensitive and accurate SERS/RRS dual-spectroscopic immunosensor for clenbuterol based on nitrogen/silver-codoped carbon dots catalytic amplification. *Talanta* **2020**, *209*, 120529. [CrossRef]
90. Abdussalam, A.; Xu, G. Recent advances in electrochemiluminescence luminophores. *Anal. Bioanal. Chem.* **2022**, *414*, 131–146. [CrossRef] [PubMed]
91. Dong, T.T.; Tang, Q.H.; Chen, M.; Deng, A.P.; Li, J.G. Ultrasensitive Electrochemiluminescent competitive immunoassay for β-Adrenergic agonist salbutamol based on quantum dots and enzymatic amplification. *J. Electrochem. Soc.* **2016**, *163*, 62–67. [CrossRef]
92. Cai, F.; Wang, N.; Dong, T.; Deng, A.; Li, J. Dual-signal amplified electrochemiluminescence immunoassay for salbutamol based on quantum dots and gold nanoparticle-labeled horseradish peroxidase. *Analyst* **2015**, *140*, 5885–5890. [CrossRef]
93. Yan, Q.N.; Cai, F.D.; Zhao, K.; Wu, K.; Deng, A.P.; Li, J.G. Dual-signal amplified electrochemiluminescent immunosensor for detecting ractopamine using quantum dots and gold nanoparticles-labeled horseradish peroxidase. *ECS J. Solid State Sci. Technol.* **2017**, *6*, 56–62. [CrossRef]
94. Dong, T.T.; Tang, Q.H.; Zhao, K.; Deng, A.P.; Li, J.G. Ultrasensitive electrochemiluminescent salbutamol immunoassay with dual-signal amplification using CdSe@SiO2 as label and gold nanoparticles as substrate. *Microchim. Acta* **2017**, *184*, 961–968. [CrossRef]
95. Dong, T.; Hu, L.; Zhao, K.; Deng, A.; Li, J. Multiple signal amplified electrochemiluminescent immunoassay for brombuterol detection using gold nanoparticles and polyamidoamine dendrimers-silver nanoribbon. *Anal. Chim. Acta* **2016**, *945*, 85–94. [CrossRef] [PubMed]
96. Zhu, Q.; Cai, F.; Zhang, J.; Zhao, K.; Deng, A.; Li, J. Highly sensitive electrochemiluminescent immunosensor based on gold nanoparticles-functionalized zinc oxide nanorod and poly(amidoamine)-graphene for detecting brombuterol. *Biosens. Bioelectron.* **2016**, *86*, 899–906. [CrossRef] [PubMed]

97. Zhu, Q.; Liu, H.X.; Zhang, J.; Wu, K.; Deng, A.P.; Li, J.G. Ultrasensitive QDs based electrochemiluminescent immunosensor for detecting ractopamine using AuNPs and Au nanoparticles@PDDA-graphene as amplifier. *Sens. Actuators B* **2017**, *243*, 121–129. [CrossRef]
98. Li, P.; Ma, G.; Wu, K.; Deng, A.; Li, J. An electrochemiluminescence energy resonance transfer system for highly sensitive detection of brombuterol. *Talanta* **2021**, *223*, 121687. [CrossRef]
99. Lin, C.H.; Lin, M.J.; Huang, J.D.; Chuang, Y.S.; Kuo, Y.F.; Chen, J.C.; Wu, C.C. Label-free impedimetric immunosensors modulated by protein A/bovine serum albumin layer for ultrasensitive detection of salbutamol. *Sensors* **2020**, *20*, 771. [CrossRef]
100. He, L.H.; Guo, C.P.; Song, Y.P.; Zhang, S.; Wang, M.H.; Peng, D.L.; Fang, S.M.; Zhang, Z.H.; Liu, C.S. Chitosan stabilized gold nanoparticle based electrochemical ractopamine immunoassay. *Microchim. Acta* **2017**, *184*, 2919–2924. [CrossRef]
101. Zhan, B.L.; Zhang, Y.T.; Zhao, X. High sensitive sol-gel based electrochemical immunosensor for Clenbuterol Determination. *Int. J. Electrochem. Sci.* **2021**, *16*, 211124. [CrossRef]
102. Cui, Z.T.; Cai, Y.Y.; Wu, D.; Yu, H.Q.; Li, Y.; Mao, K.X.; Wang, H.; Fan, H.X.; Wei, Q.; Du, B. An ultrasensitive electrochemical immunosensor for the detection of salbutamol based on Pd@SBA-15 and ionic liquid. *Electrochim. Acta* **2012**, *69*, 79–85. [CrossRef]
103. Wang, H.; Zhang, Y.; Li, H.; Du, B.; Ma, H.; Wu, D.; Wei, Q. A silver-palladium alloy nanoparticle-based electrochemical biosensor for simultaneous detection of ractopamine, clenbuterol and salbutamol. *Biosens. Bioelectron.* **2013**, *49*, 14–19. [CrossRef] [PubMed]
104. Gu, C.; Ren, P.; Zhang, F.; Zhao, G.; Shen, J.; Zhao, B. Detection of six β-agonists by three multiresidue immunosensors based on an anti-bovine serum albumin-ractopamine-clenbuterol-salbutamol antibody. *ACS Omega* **2020**, *5*, 5548–5555. [CrossRef] [PubMed]
105. Huang, J.D.; Lin, Q.; Zhang, X.M.; He, X.R.; Xing, X.R.; Lian, W.J.; Zuo, M.M.; Zhang, Q.Q. Electrochemical immunosensor based on polyaniline/poly (acrylic acid) and Au-hybrid graphene nanocomposite for sensitivity enhanced detection of salbutamol. *Food Res. Int.* **2011**, *44*, 92–97. [CrossRef]
106. Liu, S.; Lin, Q.; Zhang, X.M.; He, X.R.; Xing, X.R.; Lian, W.J.; Huang, J.D. Electrochemical immunosensor for salbutamol detection based onCS-Fe3O4-PAMAM-GNPs nanocomposites and HRP-MWCNTs-Ab bioconjugates for signal amplification. *Sens. Actuators B* **2011**, *156*, 71–78. [CrossRef]
107. Han, X. Development of electrochemical immunosensor for detecting sSalbutamol by competitive immune strategy. *Int. J. Electrochem. Sci.* **2020**, *15*, 7337–7346. [CrossRef]
108. Regiart, M.; Fernández-Baldo, M.A.; Spotorno, V.G.; Bertolino, F.A.; Raba, J. Ultra sensitive microfluidic immunosensor for determination of clenbuterol in bovine hair samples using electrodeposited gold nanoparticles and magnetic micro particles as bio-affinity platform. *Biosens. Bioelectron.* **2013**, *41*, 211–217. [CrossRef]
109. Talib, N.A.A.; Salam, F.; Sulaiman, Y. Development of Highly sensitive immunosensor for clenbuterol detection by using poly(3,4-ethylenedioxythiophene)/graphene oxide modified screen-printed carbon electrode. *Sensors* **2018**, *18*, 4324. [CrossRef]
110. Lai, Y.; Bai, J.; Shi, X.; Zeng, Y.; Xian, Y.; Hou, J.; Jin, L. Graphene oxide as nanocarrier for sensitive electrochemical immunoassay of clenbuterol based on labeling amplification strategy. *Talanta* **2013**, *107*, 176–182. [CrossRef]
111. Dou, Y.Z.; Jiang, Z.N.; Deng, W.P.; Su, J.; Chen, S.X.; Song, H.Y.; Aldalbahi, A.; Zuo, X.L.; Song, X.P.; Shi, J.Y.; et al. Portable detection of clenbuterol using a smartphone-based electrochemical biosensor with electric field-driven acceleration. *J. Electroanal. Chem.* **2016**, *781*, 339–344. [CrossRef]
112. Sano, T.; Smith, C.L.; Cantor, C.R. Immuno-PCR: Very sensitive antigen detection by means of specific antibody-DNA conjugates. *Science* **1992**, *258*, 120–122. [CrossRef]
113. Lei, Y.; Li, X.; Akash, M.S.; Zhou, L.; Tang, X.; Shi, W.; Liu, Z.; Chen, S. Development of analytical method for ultrasensitive detection of salbutamol utilizing DNA labeled-immunoprobe. *J. Pharm. Biomed. Anal.* **2015**, *107*, 204–208. [CrossRef] [PubMed]
114. Zhao, L.; Zhou, H.; Sun, T.; Liu, W.; He, H.; Ning, B.; Li, S.; Peng, Y.; Han, D.; Zhao, Z.; et al. Complete antigen-bridged DNA strand displacement amplification immuno-PCR assay for ultrasensitive detection of salbutamol. *Sci. Total Environ.* **2020**, *748*, 142330. [CrossRef] [PubMed]
115. Li, C.; Li, J.; Jiang, W.; Zhang, S.; Shen, J.; Wen, K.; Wang, Z. Development and application of a gel-based immunoassay for the rapid screening of salbutamol and ractopamine residues in pork. *J. Agric. Food Chem.* **2015**, *63*, 10556–10561. [CrossRef] [PubMed]
116. Zvereva, E.A.; Shpakova, N.A.; Zherdev, A.V.; Kiu, L.; Xu, C.; Eremin, S.A.; Dzantiev, B.B. Fluorescence polarization immunoassay of ractopamine. *Appl. Biochem. Microbiol.* **2016**, *52*, 673–678. [CrossRef]
117. Dong, B.; Zhao, S.; Li, H.; Wen, K.; Ke, Y.; Shen, J.; Zhang, S.; Shi, W.; Wang, Z. Design, synthesis and characterization of tracers and development of a fluorescence polarization immunoassay for the rapid detection of ractopamine in pork. *Food Chem.* **2019**, *271*, 9–17. [CrossRef]
118. Zuo, P.; Zhang, Y.; Liu, J.; Ye, B.C. Determination of beta-adrenergic agonists by hapten microarray. *Talanta* **2010**, *82*, 61–66. [CrossRef]
119. Kong, J.; Jiang, L.; Su, X.; Qin, J.; Du, Y.; Lin, B. Integrated microfluidic immunoassay for the rapid determination of clenbuterol. *Lab Chip.* **2009**, *9*, 1541–1547. [CrossRef]

foods MDPI

Article

Cascade-Enhanced Lateral Flow Immunoassay for Sensitive Detection of Okadaic Acid in Seawater, Fish, and Seafood

Olga D. Hendrickson, Elena A. Zvereva, Anatoly V. Zherdev and Boris B. Dzantiev *

Bach Institute of Biochemistry, Research Center of Biotechnology of the Russian Academy of Sciences, Leninsky Prospect 33, 119071 Moscow, Russia; odhendrick@gmail.com (O.D.H.); zverevaea@yandex.ru (E.A.Z.); zherdev@inbi.ras.ru (A.V.Z.)
* Correspondence: dzantiev@inbi.ras.ru; Tel.: +7-495-954-31-42

Abstract: In this investigation, a new approach for developing a sensitive lateral flow immunoassay (LFIA) was proposed for the detection of the hazardous marine toxin okadaic acid (OA). It is based on the indirect format with anti-species antibodies labeled by gold nanoparticles (AuNPs) and cascade signal amplification. The latter is performed by first passing a mixture of anti-OA antibodies and a tested sample along the immunochromatographic test strip and then performing several cycles of the interaction of anti-species antibodies conjugated with AuNPs with free antibodies, which bind to anti-species antibodies but are not specific to the target analyte. As a result, branched aggregates are formed, due to which the colorimetric signal intensification occurs. The developed test system enabled the detection of OA with an instrumental detection limit of 30 pg/mL and a cutoff of 1 ng/mL, which exceeds these characteristics in the LFIA without amplification by 7 and 2 times, respectively. The OA recoveries from seawater, fish, and seafood varied from 76.9% to 126%. The test system may be required for point-of-care monitoring of samples for phycotoxin contamination; the developed principle of signal amplification can be used in cases where highly sensitive detection of trace amounts of a contaminant is required.

Keywords: phycotoxins; okadaic acid; lateral flow immunoassay; signal amplification; seawater; seafood

Citation: Hendrickson, O.D.; Zvereva, E.A.; Zherdev, A.V.; Dzantiev, B.B. Cascade-Enhanced Lateral Flow Immunoassay for Sensitive Detection of Okadaic Acid in Seawater, Fish, and Seafood. *Foods* **2022**, *11*, 1691. https://doi.org/10.3390/foods11121691

Academic Editor: Maojun Jin

Received: 16 May 2022
Accepted: 6 June 2022
Published: 9 June 2022

Publisher's Note: MDPI stays neutral with regard to jurisdictional claims in published maps and institutional affiliations.

Copyright: © 2022 by the authors. Licensee MDPI, Basel, Switzerland. This article is an open access article distributed under the terms and conditions of the Creative Commons Attribution (CC BY) license (https://creativecommons.org/licenses/by/4.0/).

1. Introduction

Among a large number of compounds related to food contaminants, a special place belongs to phycotoxins—extremely toxic compounds produced by microalgae and cyanobacteria that are part of the plankton and benthos of the world ocean [1–3]. Normally, algae and cyanobacteria are a necessary component of aquatic ecosystems; however, under favorable conditions, these organisms actively multiply, which leads to the so-called water bloom (for example, "red tides") and changes in ecosystems [4]. Algae and cyanobacteria are an intermediate link in the food chain, serving as food for macroorganisms such as fish and shellfish. The transfer of phycotoxins to aquatic animals can lead to significant environmental and economic consequences causing their death and thereby harming fish farming and fisheries [5]. In addition, water containing phycotoxins is not suitable for drinking because most of them are thermostable, which causes difficulties for purification in the water supply. Human consumption of fish and shellfish contaminated with phycotoxins leads to poisoning, sometimes massive because phycotoxins affect health even in small concentrations [6,7].

Phycotoxins produced by several species of dinoflagellates and causing diarrheic shellfish poisoning (DSP) include, in particular, okadaic acid (OA), the mechanism of action of which is manifested in the inhibition of protein phosphatase activity [8,9]. Because OA is a lipophilic compound, it accumulates in the fatty tissues of shellfish and fish [9]. DSP, which develops almost immediately after the intake of OA-containing seafood, is characterized by nausea, vomiting, abdominal pain, and profuse diarrhea [10]. Given the high toxicity of OA, its content in food products is strictly regulated. Thus, according to the European Union regulatory limit, the content of OA in mollusk tissues should not exceed 0.16 μg/kg [11].

Ensuring food quality and safety requires the control of the contamination of raw materials, semi-finished, and finished food products. This requirement applies in particular to fish, seafood, and related foodstuffs, which due to their palatability and nutritional value are included in the diet in many countries and are the basis for standard and gourmet dishes. Therefore, phycotoxins are included in the list of mandatory controlled food contaminants, and analytical methods for their detection are an essential tool to implement their monitoring. For precise and sensitive determination of phycotoxins, complex analytical methods such as high-performance liquid chromatography–mass spectrometry are often used, which require specialized laboratories with complex and expensive equipment and highly qualified operators [12,13]. These approaches cannot provide a rapid point-of-care determination of the toxicant, especially for mass screening of samples. From this point of view, immunochemical methods, particularly the LFIA based on a combination of chromatography and highly specific interaction of analytes with antibodies, can be an alternative or addition to complex arbitrage analytical methods. It provides rapid results (10–20 min) on not only qualitative (phycotoxin presence/absence) but in many cases also quantitative (its concentration) characteristics [14–16].

It should be noted that the peculiarity of multicomponent food and water matrices is that before analysis, a sample preparation procedure often associated with multiple dilutions of samples is required. As a result, the sensitivity of the analysis developed in model conditions (determination in a buffer) can be insufficient to ensure a reliable detection in real phycotoxin-containing samples. Therefore, it is necessary to have a margin in the assay sensitivity that allows for the reliable detection of a phycotoxin in the sample. Therefore, the creation of approaches aimed at lowering the limit of detection (LOD) is an extremely popular direction in the development of analytical systems including LFIAs [17–20].

The LFIA of OA has been described in several studies [21–25]. Most of these works are based on the routine direct competitive LFIA with AuNPs as a label for specific antibodies. The reported test systems enable the determination of OA with LODs varying in the range of 0.1–50 ng/mL. Only one recent study is devoted to the development of an enhancement strategy based on the catalysis of Au@Pt nanoparticles and horseradish peroxidase [25]. With this approach, the authors achieved an OA LOD of 0.04 ng/mL. The developed LFIAs were tested for the detection of OA in real samples of shellfish.

In contrast to the studies described above, where the same principle of competitive interaction and one-stage assembling of a detectable complex are reported, in this study, a highly sensitive indirect LFIA of OA was developed based on the amplification of the analytical signal, which is provided by a cascade of interactions between gold-labeled secondary antibodies and free antibodies having specificity to these secondary antibodies but not to OA. The achieved analytical characteristics exceed those in all published studies on OA immunochromatography, including an enhanced LFIA. The developed enhanced LFIA was applied for the detection of OA in spiked samples of seawater, fish, and seafood.

2. Materials and Methods

2.1. Reagents, Materials, Equipment, and Software

OA, gold (III) chloride hydrate ($HAuCl_4 \times H_2O$), sodium azide, methanol, sucrose, Triton X-100, and bovine serum albumin (BSA) (Sigma-Aldrich, Saint Louis, MO, USA) were used. Goat anti-mouse immunoglobulins (GAMI) and donkey anti-goat immunoglobulins (DAGI) were purchased from Arista Biologicals (Allentown, PA, USA). Monoclonal antibodies (MAbs) to OA (clone 7E1) were purchased from Santa Cruz Biotechnology (Dallas, TX, USA). All other compounds were analytically pure.

For the LFIA, a CNPC-SS12 nitrocellulose membrane fixed on the plastic support and a GFB-R4 membrane (Advanced Microdevices, Ambala Cantt, India) were used as a working membrane and a sample pad, respectively. As an adsorption pad, a ReliaFlow 319 membrane (Ahlstrom-Munksjö, Helsinki, Finland) was applied.

Transmission electron microscopy (TEM) was performed on a CX-100 microscope (Jeol, Tokyo, Japan). A Zenyth 3100 vertical photometer (Anthos Labtec Instruments, Wals, Austria) was used to register the optical density (OD) of gold solutions. An Iso-Flow dispenser (Imagene Technology, Hanover, NH, USA) was utilized to apply the reagents on the immunochromatographic working membrane (at a rate of 0.1 μL per mm), and an automatic guillotine (KinBio, Shanghai, China) was used to cut it into test strips. To assess bands' coloration, a CanoScan LiDE 90 scanner (Canon, Tokyo, Japan) and TotalLab software (Nonlinear Dynamics, Newcastle upon Tyne, Great Britain) were used. Origin software (OriginLab, Northampton, MA, USA) was applied to estimate the analytical characteristics of the developed test systems.

2.2. Synthesis of AuNPs and Their Conjugation with GAMI

AuNPs were obtained by the standard approach described in [26] and characterized by TEM as reported in [27]. To determine the GAMI concentration for conjugation with AuNPs, a flocculation curve was obtained. The pH of the AuNP solution (OD_{520} = 1) was adjusted to 9.0 with 100 mM sodium carbonate. After that, AuNPs (500 μL) were mixed with GAMI solutions (0.5–200 μg/mL, 50 μL in 10 mM Tris-HCl, pH 8.5) and incubated for 10 min at room temperature. Then, 10% sodium chloride (50 μL) was added and OD_{580} was measured after stirring. Finally, the dependence of OD_{580} versus GAMI concentration was built. OD_{580} was chosen because of the changes in spectral characteristics of AuNPs solution after the aggregation of nanoparticles caused by the addition of the coagulating agent (NaCl) [28].

To obtain labeled antibodies, GAMI were added to AuNPs (OD_{520} = 1, pH 9.0) in the concentration of 6 μg/mL. The mixture was shaken for 45 min at room temperature, followed by the addition of a 10% water solution of BSA (40:1, *v*/*v*) and vigorous stirring for 15 min. Then, the GAMI–AuNPs conjugate was pelleted by centrifugation at 9500× *g* for 35 min at 4 °C. The precipitate was resuspended to an OD_{520} = 15 in 10 mM Tris buffer, pH 8.5, containing 1% BSA, 1% sucrose, and 0.1% sodium azide. The conjugate was stored at 4 °C.

2.3. Preparation of Test Strips

In the case of the standard LFIA, test strips were combined from the working membrane, a sample pad, and an adsorption pad. For the enhanced LFIA, the plastic support was cut until the lower edge of the working membrane. For both formats of the LFIA, OA–BSA (0.5 mg/mL in PBS) and DAGI (0.1 mg/mL in PBS for the standard LFIA and 0.05 mg/mL for the enhanced LFIA) were applied onto a working membrane to form a test (T) zone and a control (C) zone, respectively. The multimembrane composite was dried overnight at room temperature and for 1.5 h at 37 °C and then cut into strips of 3.0 mm width. The strips were stored at room temperature in a sealed package with silica gel.

2.4. Pretreatment of Seawater and Seafood Samples

A seawater sample was taken from the Aegean Sea (Fethiye region, Turkey) and stored at 4 °C. Before analysis, Triton X-100 was added to seawater (0.05%). Then, the obtained mixture was diluted by 10 times with PBST and spiked with known concentrations of OA.

The real seafood samples included fish (trout from the Barents Sea, Russia), tiger shrimps, and scallops (both from the Sea of Okhotsk, Russia) purchased from local food stores. For the sample preparation of fish and seafood, the following technique was used: first, samples were minced into a homogeneous mass using a household blender. Then, to a 0.5 g sample, OA (50 μL, 1 μg/mL, which corresponds to 100 ng/g), and 5 mL of the methanol–water mixture (1:1) were added. The mixtures were stirred for 5 min and centrifuged at 1500× *g* for 10 min. The supernatants were collected and stored at −18 °C. Before the LFIA, the extracts were diluted 10 times with PBST.

2.5. LFIA of OA

For the determination of OA, its solutions (15.2 pg/mL–100 ng/mL, 50 μL in PBST) were mixed with anti-OA MAbs (0.1 μg/mL, 50 μL in PBST) and GAMI–AuNPs (2.5 μL, OD_{520} = 15) and incubated for 3 min at room temperature. Then, the test strips were incubated in the obtained solutions for 15 min. To estimate the LFIA results, test strips were removed from the solutions, blotted and scanned. Then, bands' coloration in the T zone was assessed.

2.6. LFIA of OA with Cascade Signal Amplification

OA solutions (0.08 pg/mL–50 ng/mL, 10 μL in PBST) were mixed with anti-OA MAbs (0.01 μg/mL, 10 μL in PBST) and incubated for 3 min at room temperature. Then, the test strips were immersed into the mixture and incubated for 5 min. After that, the test strips were transferred to the solution of GAMI–AuNPs (2 μL in 20 μL of PBST) and incubated in it. This and all other stages were carried out for 7 min. Then, 2 cycles of signal amplification were performed. A single cycle included the following steps: the test strips were transferred to the solution of DAGI (20 μL, 500 ng/mL in PBST) and, after incubation, were transferred to the solution of GAMI–AuNPs (2 μL in 20 μL of PBST). Finally, the test strips were processed as described above.

In the case of the enhanced LFIA in real samples, spiked extracts of fish or seafood or seawater pretreated as described above were added instead of OA buffer solutions. All other stages of the analysis were the same.

2.7. Evaluation of the Assay Results

The plots of color intensity or OD (y) versus the OA concentrations (x) were built and fitted to a four-parameter logistic function using Origin software (OriginLab, Northampton, MA, USA). The LODs, cutoffs, and working ranges were evaluated as described in [28,29].

3. Results and Discussion

3.1. Obtaining the Immunoreagents

To develop the LFIA of OA, colloidal gold was used as a traditional label in immunochromatography, which is characterized by a standardized synthetic protocol, long-term stability, and a high colorimetric signal that provides sensitive and reliable analyte determination both visually and instrumentally. AuNPs were obtained through the reduction of $HAuCl_4$ with sodium citrate [26]. In this study, AuNPs with a diameter of about 30 nm were synthesized as the most optimal marker in the LFIA [30]. TEM characterization showed that the sample contained homogeneous non-aggregated particles. The average diameter of AuNPs (a sample containing 201 nanoparticles was processed) was 30.9 ± 3.4 nm with a minimum value of 23.5 nm and a maximum value of 39.2 nm; the ellipticity was 1.1 ± 0.06 (Figure 1).

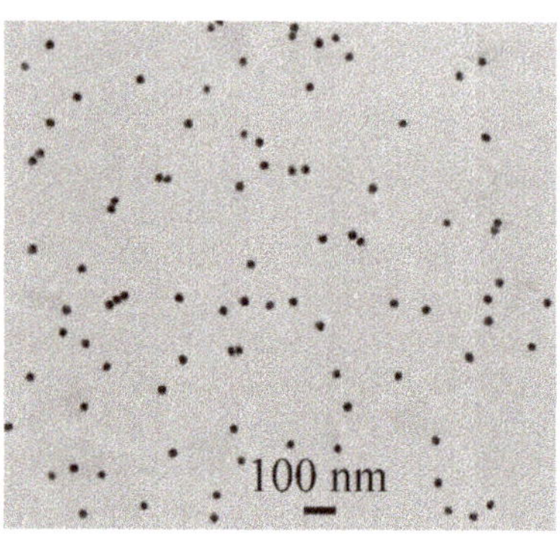

Figure 1. TEM microphotograph of AuNPs.

Both the standard and enhanced LFIAs were implemented in an indirect format implying conjugation with a label of not specific (anti-OA MAbs), but anti-species antibodies (GAMI). Before obtaining the GAMI–AuNPs conjugate, it was necessary to determine the concentration of antibodies used for complexation. This stage is very important because a correctly determined quantitative ratio of the marker and antibodies ensures the stability of the immunocomplex in media with different pH and ionic strength. The choice of GAMI concentration was carried out with the help of a flocculation curve—the dependence of the OD of the colloidal gold solution on the concentration of added antibodies in a medium with a high content of a coagulator (10% NaCl). At an insufficient concentration of antibodies, AuNPs have an unstabilized surface and are likely to aggregate, which is visualized as a change in the shade of the colloidal gold solution towards violet (a growth of the OD on the flocculation curve is observed, Figure 2).

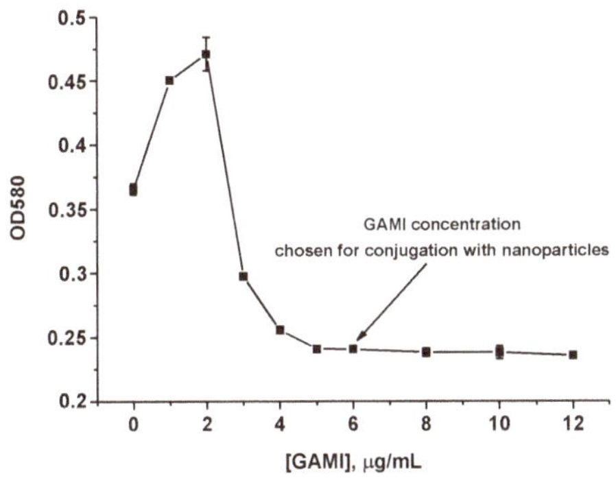

Figure 2. GAMI flocculation curve.

With an increase in protein concentration, the OD decreases reaching a plateau (a flocculation point), which indicated that the surface of AuNPs becomes steady and their aggregation stops. An antibody concentration corresponding to (or slightly above) the flocculation point is usually used to obtain a stable GAMI–AuNPs complex [27]. In our case, it corresponded to 6 μg of GAMI per 1 mL of AuNPs solution (indicated by an arrow in Figure 2).

Commercial MAbs were used as a receptor for OA. To confirm their reactivity toward OA, primary characterization (without optimization) was carried out by the indirect enzyme-linked immunosorbent analysis (ELISA), which showed that the LOD of OA was 0.5 ng/mL (Figure S1).

3.2. Standard LFIA of OA

As it was noted above, the standard LFIA was performed in the indirect format based on the competition between free OA in the sample and its protein conjugate immobilized on the immunochromatographic membrane for the binding to anti-OA MAbs. Red-colored GAMI–AuNPs were used to reveal the immune complexes formed in the test strip. A tested sample was mixed with OA-specific antibodies and GAMI–AuNPs and incubated for a short time before dipping the test strip. Then, test strips with immobilized OA–BSA in the T zone and secondary antibodies specific to GAMI in the labeled conjugate (DAGI) adsorbed in the C zone were incubated with the reaction mixture. In the presence of OA, the latter blocks specific antibodies averting the formation of the OA–BSA–MAbs–GAMI–AuNPs complex in the T zone and thereby preventing the appearance of the colored band. Contrariwise, in the absence of OA in the sample, MAbs bound with GAMI–AuNPs interact with the immobilized antigen causing the formation of a colored band. In the C zone, coloration occurs in any case owing to interaction of an excess of the labeled conjugate with DAGI adsorbed there. Thus, the intensity of the coloration in the T zone is measured to assess the concentration of OA in the sample.

For the correct comparison of the standard and enhanced variants of the LFIA, it was necessary to optimize both formats in terms of achieving the lowest possible LODs while maintaining analytical signal amplitudes sufficient for reliable determination. This was implemented by varying the assay conditions—the duration of its stages and the reagents' concentrations (Table S1). As a result, it was found that this demand was fulfilled if specific antibodies were added at a concentration of 0.1 μg/mL (at a lower amount, the intensity of the analytical signal decreased and did not meet the requirement for the assay accuracy; at a higher concentration, the LOD undesirably increased). The optimal volume of the GAMI–AuNPs conjugate added to the sample was 2.5 μL. With a smaller amount of the marker, the coloration of the zones was too pale, that is, the signal amplitude decreased; a larger amount slightly increased the brightness of the zones but the background signal and the consumption of the reagent also increased. The 3-min duration of the preincubation stage was chosen as sufficient for the progress of homogeneous immune reactions. The incubation of the test strip with the sample was carried out for 15 min, the time sufficient for the lateral flow of the reaction mixture along the membrane carriers and the implementation of all required interactions.

The LFIA optimization allowed achieving high analytical parameters: the instrumental LOD of OA was 0.2 ng/mL and the working range of the detectable concentrations was 0.31–1.3 ng/mL. Visual LOD (cutoff) was 2 ng/mL. The OA calibration curve and the test strips corresponding to concentrations plotted on the curve are shown in Figure 3. According to the obtained data, the signal amplitude reached about 2500 relative units (RU).

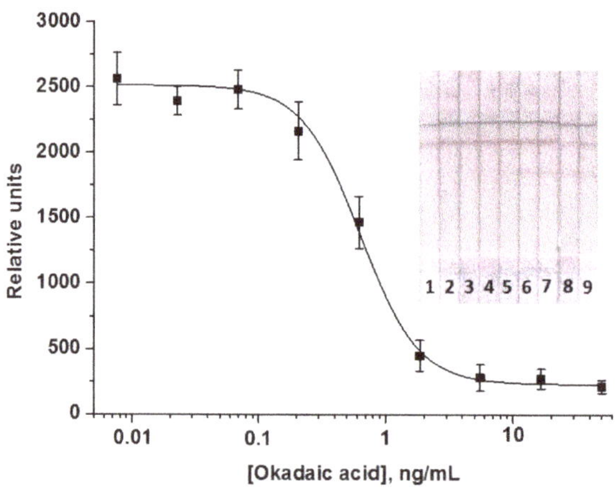

Figure 3. Calibration curve of OA in the LFIA ($n = 3$) and the corresponding test strips. Concentrations of OA were 50 ng/mL (1); 16.7 ng/mL (2); 5.6 ng/mL (3); 1.9 ng/mL (4); 0.62 ng/mL (5); 0.21 ng/mL (6); 69 ng/mL (7); 23 pg/mL (8); 7.6 pg/mL (9).

3.3. LFIA of OA with Cascade Signal Amplification

The proposed method of signal amplification is performed through a cascade of immunochemical reactions occurring on the test strip, which leads to the progressive increase in the intensity of zones' coloration [31] (see the scheme of the enhanced LFIA in Figure 3). It consists of passing a solution of specific MAbs mixed with an antigen-containing sample along the test strip followed by several (at least two) cycles of successive passing gold-labeled anti-species antibodies (GAMI–AuNPs) and free anti-GAMI antibodies that are not specific to OA (DAGI). The result of these processes is the formation of aggregates with a (GAMI–AuNPs − DAGI) × n structure, where n is the number of GAMI–AuNPs/DAGI passing cycles. Theoretically, the number of such cycles (and, accordingly, the number of layers formed in the T zone) is unlimited, which eliminates stoichiometric restrictions on the amount of the markers attached to one immunoreagent molecule immobilized on the membrane.

Under cascade amplification, the stages of specific interaction with the antigen in the sample and the introduction of a colored label into the detected complex were separated. During the first reaction, the MAbs–OA–BSA complex is formed in the analytical zone. For its detection, a second reaction is performed: a solution of GAMI–AuNPs is passed along the test strip, which leads to their binding in the T zone and the appearance of a colored band. It should be noted that the AuNPs are evenly coated with anti-species antibodies. Hence, if binding with the MAbs–OA–BSA complex occurs on one side of the nanoparticle, the opposite side remains free for other interactions. Therefore, it is possible to carry out additional interactions that increase the incorporation of the marker into the immune complex. For this purpose, a solution of free antibodies (DAGI, not specific to the target analyte but binding to anti-species antibodies that are already included in the complex on the membrane) is passed along the test strip. Then, a solution of GAMI–AuNPs is again passed to label the resulting new layer of immunoglobulins. As a result, a complex multilayer structure is formed in the T zone, where one antigen–antibody complex induces the binding of a large amount of colored marker (Figure 4). The separation of stages enables independent control of the content of specific antibodies and a colored marker in the system, which, in turn, allows for increasing the color intensity of the T zone and thereby reducing the LOD of the analyte.

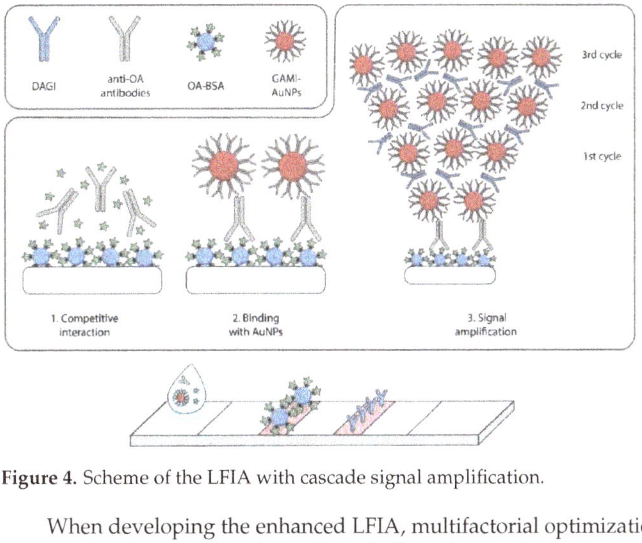

Figure 4. Scheme of the LFIA with cascade signal amplification.

When developing the enhanced LFIA, multifactorial optimization was carried out because such a format differed significantly from the standard one and required the selection of many assay parameters. First, it was necessary to establish the number of amplification cycles that have to be included in the analysis taking into account that the aim of the development is the maximum decrease in the LOD of OA. According to the principle described above, it is achieved by increasing the number of amplification cycles. However, a large number of the assay stages will prolong its total duration. Regarding this, it was necessary to choose such detection conditions under which the achievement of high analytical parameters of the test system does not contradict the rapid detection as the basic advantage of the LFIA.

The enhanced LFIA of OA was initially carried out under the same conditions as the standard one, i.e., with the same composition of test strips, the volume of the analyzed sample (100 µL), the concentration of reagents deposited on the working membrane, specific antibodies (0.1 µg/mL), and GAMI–AuNPs in solution (2.5 µL for each cycle). Under these conditions, one or more amplification cycles were implemented by sequentially as test strips were incubated in GAMI–AuNPs and DAGI solutions. As a result of the experiments, several important features were revealed.

First, at least 12–15 min duration of the incubation of full-size test strips with a 100 µL reaction mixture is required to ensure maximum liquid absorption and effectiveness of immune interactions on the membrane. Consequently, the detection with 1–3 amplification cycles takes approximately 50–100 min, which is comparable to the microplate ELISA and seriously contravenes the LFIA rapidity. The total time for the liquid movement along the test strip consists of the durations of movements along its components, including a sample pad. Therefore, we decided to cut the test strips up to the lower edge of the working membrane, thus reducing the length of the strip by about one-third. In this case, the sample volume of 100 µL also became excessive and was reduced. As a result of the experiments on the selection of the combination of the volume of the reaction mixture and the optimal time of its incubation with the test strip, the 20 µL sample volume (we varied it in the range of 15–30 µL) and 7 min incubation time (we varied in the range of 5–10 min) were chosen. For the initial incubation of a dry test strip with OA and MAbs-containing sample, 5 min was sufficient to completely absorb a reaction mixture. Therefore, the total assay duration was 29 min for one cycle and increased by 14 min for each subsequent cycle. In this case, it was irrational to carry out more than 2–3 amplification cycles from the point of view of rapid detection.

Secondly, when using the same concentration of anti-OA MAbs as in the standard LFIA, the intensity of the colorimetric signal in the enhanced LFIA significantly increased (up to >10,000 RU) depending on the number of amplification cycles (Figure 5).

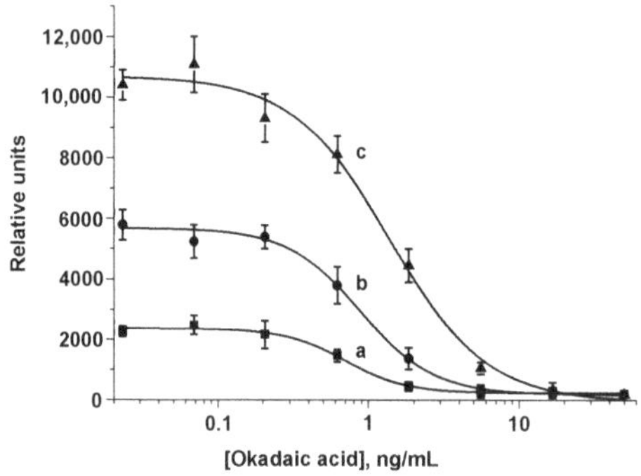

Figure 5. Calibration curves of OA in the standard LFIA (**a**) and in the enhanced LFIA with 2 (**b**) and 3 (**c**) amplification cycles (n = 3).

For a correct comparison with the standard LFIA, it was advisable to unify the colorimetric signal generated as a result of detection (~2500 RU). It was achieved by reducing the concentration of specific antibodies, which, in turn, decreased the OA LOD. In the enhanced LFIA, the concentration of MAbs was reduced by an order of magnitude (to 10 ng/mL, we varied it in the range from 8 to 40 ng/mL). It should be noted that under this concentration of MAbs, the coloration in the T zone is absent in the standard LFIA. At the same time, because the number of incubations with GAMI–AuNPs increases by at least one (with a single amplification cycle) and at each stage the coloration in the zones becomes brighter, it becomes possible to reduce the volume of the added GAMI–AuNPs conjugate. Finally, 2 µL of the GAMI–AuNPs conjugate was added at each amplification stage. On the one hand, this is slightly less than the volume added in the standard LFIA, but on the other hand, due to a significant decrease in the concentration of anti-OA MAbs, the increase in the signal occurs gradually precisely during the cascade amplification. Another parameter for optimization was the DAGI concentration. The concentration of 500 ng/mL was chosen as optimal (we varied it in the range of 150–1000 ng/mL). At a lower amount of DAGIs, the enhancement of zones' coloration during the cascade was not sufficiently pronounced; at a higher amount, the background signal increased significantly. Because during cascade amplification, an increase in the signal also occurred in the C zone of the test strip, the DAGI concentration there was 2 times reduced compared to the standard LFIA (to 0.05 ng/mL). The concentration of the OA–BSA conjugate in the T zone was unchanged (0.5 ng/mL).

Under the conditions chosen during optimization, the LFIAs with one, two, and three amplification cycles were implemented. With a single amplification cycle, the amplitude of the analytical signal was too low (<2000 RU), which negatively affected the reliability and accuracy of the analysis. Therefore, this variant was excluded from the comparison, and only two- and three-cascade formats were considered. It was demonstrated that at comparable instrumental LODs, the visual LOD (cutoff) after three amplifications increased. This can be explained by the growth of the background signal as a result of non-specific interactions that may occur as a result of a significant increase in the incubation time of the test strip during multiple steps of the three-stage amplification. In addition, the total assay duration for three cascades was 57 min. Thus, the LFIA with two cycles of signal amplification was chosen. The OA calibration curve obtained in the optimized LFIA with cascade amplification is shown in Figure 6. The OA LOD was 0.03 ng/mL, which is almost 7 times lower than in the standard LFIA; the working range of the detectable concentrations was 0.07–0.85 ng/mL. Cutoff decreased by 2 times—down to 1 ng/mL. The obtained results allowed for testing the developed analysis for detecting OA in real samples.

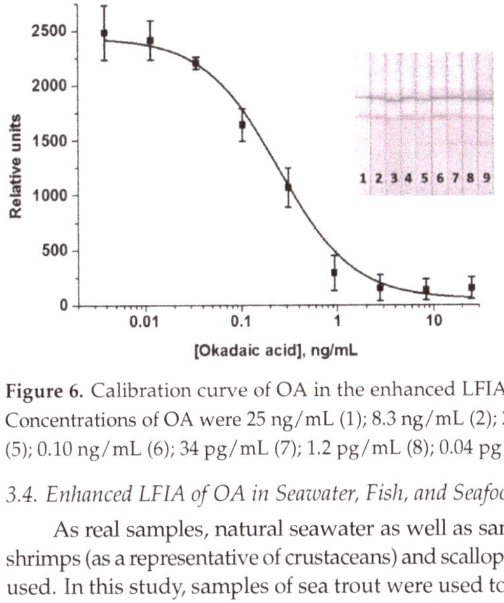

Figure 6. Calibration curve of OA in the enhanced LFIA (n = 3) and the corresponding test strips. Concentrations of OA were 25 ng/mL (1); 8.3 ng/mL (2); 2.8 ng/mL (3); 0.93 ng/mL (4); 0.31 ng/mL (5); 0.10 ng/mL (6); 34 pg/mL (7); 1.2 pg/mL (8); 0.04 pg/mL (9).

3.4. Enhanced LFIA of OA in Seawater, Fish, and Seafood

As real samples, natural seawater as well as samples of fish (trout) and seafood—tiger shrimps (as a representative of crustaceans) and scallops (as a representative of mollusks)—were used. In this study, samples of sea trout were used to develop a method for sample preparation and further LFIA. Freshwater trout can also be contaminated with DSP toxins produced by dinoflagellates, which are the important group of phytoplankton in marine and fresh waters [32]. Rapid and simple sample preparation techniques were proposed. Triton X-100 detergent was added to seawater to increase the mobility of the reaction mixture along the immunochromatographic membrane and decrease the non-specific binding. Then, seawater was diluted with PBST. For fish and seafood, methanol:water extraction was applied, followed by the dilution of the extracts with PBST.

Using it, both blank and spiked samples were processed. In the latter case, sample preparation was aimed not only at reducing the matrix effect on the results of testing, but also at effective OA extraction. Confirmation of the OA absence in seawater, fish, and seafood (before spiking) was carried out using OA ELISA kits (EuroProxima, Arnhem, the Netherlands). OA recovery values calculated on the base of the OA calibration curve in PBST are presented in Table 1.

Table 1. Recoveries of OA from seawater, fish, and seafood (n = 3).

Seawater						
Added OA, ng/mL		Detected OA ± SD [1] (ng/mL)			Recovery ± SD (%)	
0.5		0.45 ± 0.04			89.2 ± 8.1	
0.75		0.62 ± 0.04			82.0 ± 4.7	
Fish and Seafood						
Added OA, ng/mL	Detected OA ± SD (ng/g)/	Recovery ± SD (%)	Detected OA ± SD (ng/g)/	Recovery ± SD (%)	Detected OA ± SD (ng/g)/	Recovery ± SD (%)
	Trout		Shrimps		Scallops	
50	38.5 ± 1.0	76.9 ± 1.9	47.8 ± 5.7	95.60 ± 11.4	51.9 ± 2.1	103.7 ± 4.2
100	113.9 ± 10	113.9 ± 10	123 ± 1.2	123 ± 1.2	126 ± 15	126 ± 15

[1] SD—standard deviation, n = 3.

The data in Table 1 indicate that the developed LFIA enables the determination of 76.9–126% OA in seawater, fish, and seafood samples. The accuracy of the developed assay was confirmed using the OA ELISA kits (EuroProxima, Arnhem, the Netherlands). High correlation coefficients (0.985, n = 10) between the amounts of OA determined by the LFIA and the ELISA were demonstrated.

3.5. Advantages of the Developed Test System

Comparison with previously reported studies on the LFIA of OA demonstrates that the achieved LOD (30 pg/mL) and cutoff (1 ng/mL) are the minimum among all reported studies even that where amplification approach is used (Table 2). The assay sensitivity is much beyond the official requirements, which allows varying sample preparation conditions, including significant dilution of complex matrix samples. In addition, the proposed amplification approach is very simple and does not require the synthesis of any new immunoreagents and labeled complexes: the enhanced LFIA includes the same set of reagents as the conventional indirect competitive LFIA. As additional advantages of the developed test system, a very small volume of the test sample (20 µL) and extremely low consumption of specific antibodies may be noted.

Table 2. Comparison of the developed LFIA with previous studies in this field.

LFIA Format	Label	LOD, ng/mL	Cutoff, ng/mL	Detected Real Samples	Reference
Direct competitive	Anti-OA MAbs—AuNPs	10	50	Shellfish	[23]
Direct competitive	Anti-OA MAbs—AuNPs	3.12	6.25	Mussels	[20]
Direct competitive	Anti-OA MAbs—AuNPs	n/p [1]	5	Clams, scallops, mussels, and oysters	[22]
Direct competitive	Anti-OA MAbs—AuNPs	100	1	Shellfish	[21]
Catalysis enhancement	Anti-OA MAbs—Au@PtNPs	0.04	n/p	Oysters, mussels, and clams	[24]
Indirect competitive	GAMI–AuNPs	0.03	1	Seawater, fish (trout), tiger shrimps, and scallops	This study

[1] Not presented.

Unlike previous works, in our investigation, the analysis of real matrices was not limited only to seafood. We enlarged a panel of tested samples to seawater as a primary target for contamination of phycotoxins and representatives of fish (trout), mollusks (scallops), and crustaceans (tiger shrimps), which additionally confirmed the competitive capabilities of the test system. The sample processing procedure was very simple and, most importantly, short: sample preparation of fish and seafood took only 20 min, and for seawater, it was reduced to a simple dilution with a buffer.

4. Conclusions

An approach for a very sensitive determination of OA as a hazardous marine toxin was developed. It was based on the indirect lateral flow immunoassay with signal amplification carried out due to the introduction of a large amount of marker into the test system. Several successive cycles (a cascade) of the interaction with gold-labeled anti-species antibodies ensure the formation of branched structures, thus significantly increasing the amount of the marker attached to the initial immune complex. The LOD, cutoff, and linear range of the test system were 0.03, 1, and 0.07–0.85 ng/mL, respectively. The LFIA was successfully applied to detect OA in spiked samples of seawater, fish, shrimps, and scallops with the recoveries of 76.9% to 126%, which confirms the promise of this method for the sensitive detection of phycotoxins in various objects. A comparison of the developed cascade-enhanced lateral flow immunoassay with already used methods demonstrated that this method required less time and successfully detected OA.

Supplementary Materials: The following supporting information can be downloaded at: https://www.mdpi.com/article/10.3390/foods11121691/s1, Figure S1: Calibration curve of OA in the indirect ELISA; Table S1: Parameters varied during optimization.

Author Contributions: The manuscript was written through contributions of all authors. O.D.H.: methodology, investigation, formal analysis, writing—original draft, data curation, writing—review and editing; E.A.Z.: investigation, data curation; A.V.Z.: conceptualization, writing—review and editing, project administration; B.B.D.: writing—review and editing, project administration, supervision. All authors have read and agreed to the published version of the manuscript.

Funding: This research was funded by the Russian Science Foundation, grant number 20-43-07001.

Institutional Review Board Statement: Not applicable.

Informed Consent Statement: Not applicable.

Data Availability Statement: The original contributions presented in the study are included in the article and Supplementary Materials; further inquiries can be directed to the corresponding author.

Acknowledgments: The authors are grateful to A.E. Urusov for useful discussions, S.M. Pridvorova (Research Center of Biotechnology) for TEM studies, and D.S. Popravko for the design of Figure 4.

Conflicts of Interest: The authors declare no conflict of interest.

References

1. Farabegoli, F.; Blanco, L.; Rodriguez, L.P.; Vieites, J.M.; Cabado, A.G. Phycotoxins in marine shellfish: Origin, occurrence and effects on humans. *Mar. Drugs* **2018**, *16*, 188. [CrossRef] [PubMed]
2. Louzao, M.C.; Vilariño, N.; Vale, C.; Costas, C.; Cao, A.; Raposo-Garcia, S.; Vieytes, M.R.; Botana, L.M. Current trends and new challenges in marine phycotoxins. *Mar. Drugs* **2022**, *20*, 198. [CrossRef] [PubMed]
3. Nicolas, J.; Hoogenboom, R.L.A.P.; Hendriksen, P.J.M.; Bodero, M.; Bovee, T.F.H.; Rietjens, I.M.C.M.; Gerssen, A. Marine biotoxins and associated outbreaks following seafood consumption: Prevention and surveillance in the 21st century. *Glob. Food Secur.-Agric. Policy Econ. Environ.* **2017**, *1*, 11–21. [CrossRef]
4. Neves, R.A.F.; Nascimento, S.M.; Santos, L.N. Harmful algal blooms and shellfish in the marine environment: An overview of the main molluscan responses, toxin dynamics, and risks for human health. *Environ. Sci. Pollut. Res.* **2021**, *28*, 55846–55868. [CrossRef] [PubMed]
5. Corriere, M.; Solino, L.; Costa, P.R. Effects of the marine biotoxins okadaic acid and dinophysistoxins on fish. *J. Mar. Sci. Eng.* **2021**, *9*, 293. [CrossRef]
6. Vilarino, N.; Louzao, M.C.; Abal, P.; Cagide, E.; Carrera, C.; Vieytes, M.R.; Botana, L.M. Human poisoning from marine toxins: Unknowns for optimal consumer protection. *Toxins* **2018**, *10*, 324. [CrossRef]
7. Morabito, S.; Silvestro, S.; Faggio, C. How the marine biotoxins affect human health. *Nat. Prod. Res.* **2018**, *32*, 621–631. [CrossRef]
8. Fu, L.L.; Zhao, X.Y.; Ji, L.D.; Xu, J. Okadaic acid (OA): Toxicity, detection and detoxification. *Toxicon* **2019**, *160*, 1–7. [CrossRef]
9. Valdiglesias, V.; Prego-Faraldo, M.V.; Pasaro, E.; Mendez, J.; Laffon, B. Okadaic acid: More than a diarrheic toxin. *Mar. Drugs* **2013**, *11*, 4328–4349. [CrossRef]
10. Franchini, A.; Malagoli, D.; Ottaviani, E. Targets and effects of yessotoxin, okadaic acid and palytoxin: A differential review. *Mar. Drugs* **2010**, *8*, 658–677. [CrossRef]
11. Marine biotoxins in shellfish—Summary on regulated marine biotoxins. Scientific opinion of the panel on contaminants in the food chain in feed and food. *EFSA J.* **2009**, *1306*, 1–23. [CrossRef]
12. Daguer, H.; Hoff, R.B.; Molognoni, L.; Kleemann, C.R.; Felizardo, L.V. Outbreaks, toxicology, and analytical methods of marine toxins in seafood. *Curr. Opin. Food Sci.* **2018**, *24*, 43–55. [CrossRef]
13. Rodriguez, I.; Vieytes, M.R.; Alfonso, A. Analytical challenges for regulated marine toxins. Detection methods. *Curr. Opin. Food Sci.* **2017**, *18*, 29–36. [CrossRef]

14. Di Nardo, F.; Chiarello, M.; Cavalera, S.; Baggiani, C.; Anfossi, L. Ten years of lateral flow immunoassay technique applications: Trends, challenges and future perspectives. *Sensors* **2021**, *21*, 5185. [CrossRef]
15. Dillon, M.; Zaczek-Moczydlowska, M.A.; Edwards, C.; Turner, A.D.; Miller, P.I.; Moore, H.; McKinney, A.; Lawton, L.; Campbell, K. Current trends and challenges for rapid smart diagnostics at point-of-site testing for marine toxins. *Sensors* **2021**, *21*, 2499. [CrossRef]
16. Anfossi, L.; Baggiani, C.; Giovannoli, C.; D'Arco, G.; Giraudi, G. Lateral-flow immunoassays for mycotoxins and phycotoxins: A review. *Anal. Bioanal. Chem.* **2013**, *405*, 467–480. [CrossRef]
17. Liu, Y.L.; Zhan, L.; Qin, Z.P.; Sackrison, J.; Bischof, J.C. Ultrasensitive and highly specific lateral flow assays for point-of-care diagnosis. *ACS Nano* **2021**, *15*, 3593–3611. [CrossRef]
18. Panferov, V.G.; Safenkova, I.V.; Zherdev, A.V.; Dzantiev, B.B. Methods for increasing sensitivity of immunochromatographic test systems with colorimetric detection (review). *Appl. Biochem. Microbiol.* **2021**, *57*, 143–151. [CrossRef]
19. Shirshahi, V.; Liu, G.Z. Enhancing the analytical performance of paper lateral flow assays: From chemistry to engineering. *TrAC-Trends Anal. Chem.* **2021**, *136*, 116200. [CrossRef]
20. Zherdev, A.V.; Dzantiev, B.B. Ways to reach lower detection limits in lateral flow immunoassays. In *Rapid Test—Advances in Design, Format and Diagnostic Applications*; Anfossi, L., Ed.; InTechOpen: London, UK, 2018; pp. 9–43.
21. Hu, L.; Liu, J.; Wang, Q.; Zhang, Y.; Jia, R.; Cai, C.; Wu, W.; Chen, S.-F. Development of an immunochromatographic strip test for the rapid detection of okadaic acid in shellfish sample. *J. Appl. Phycol.* **2013**, *25*, 1091–1099. [CrossRef]
22. Wang, R.; Zeng, L.; Yang, H.; Zhong, Y.; Wang, J.; Ling, S.; Farhan Saeed, A.; Yuan, J.; Wang, S. Detection of okadaic acid (OA) using ELISA and colloidal gold immunoassay based on monoclonal antibody. *J. Hazard Mater.* **2017**, *339*, 54–160. [CrossRef]
23. Liu, B.-H.; Hung, C.-T.; Lu, C.-C.; Chou, H.-N.; Yu, F.-Y. Production of monoclonal antibody for okadaic acid and its utilization in an ultrasensitive enzyme-linked immunosorbent assay and one-step immunochromatographic strip. *J. Agric. Food Chem.* **2014**, *62*, 1254–1260. [CrossRef] [PubMed]
24. Lu, S.-Y.; Lin, C.; Li, Y.-S.; Zhou, Y.; Meng, X.-M.; Yu, S.-Y.; Li, Z.-H.; Li, L.; Ren, H.-L.; Liu, Z.-S. A screening lateral flow immunochromatographic assay for on-site detection of okadaic acid in shellfish products. *Anal. Biochem.* **2012**, *422*, 59–65. [CrossRef] [PubMed]
25. Tian, Y.Q.; Yuan, L.; Zhang, M.; He, Y.F.; Lin, X.C. Sensitive detection of the okadaic acid marine toxin in shellfish by Au@Pt NPs/horseradish peroxidase dual catalysis immunoassay. *Anal. Meth.* **2022**, *14*, 1261–1267. [CrossRef]
26. Frens, G. Controlled nucleation for the regulation of the particle size in monodisperse gold suspensions. *Nat. Phys. Sci.* **1973**, *241*, 20–22. [CrossRef]
27. Hermanson, G.T. *Bioconjugate Techniques*, 3rd ed.; Pierce Biotechnology; Thermo Fisher Scientific: Rockford, IL, USA, 2013.
28. Hendrickson, O.D.; Zvereva, E.A.; Shanin, I.A.; Zherdev, A.V.; Dzantiev, B.B. Development of a multicomponent immunochromatographic test system for the detection of fluoroquinolone and amphenicol antibiotics in dairy products. *J. Sci. Food Agric.* **2019**, *99*, 3834–3842. [CrossRef]
29. Uhrovcik, J. Strategy for determination of LOD and LOQ values–some basic aspects. *Talanta* **2014**, *119*, 178–180. [CrossRef]
30. Huang, X.; Aguilar, Z.P.; Xu, H.; Lai, W.; Xiong, Y. Membrane-based lateral flow immunochromatographic strip with nanoparticles as reporters for detection: A review. *Biosens. Bioelectron.* **2016**, *75*, 166–180. [CrossRef]
31. Urusov, A.E.; Petrakova, A.V.; Gubaidullina, M.K.; Zherdev, A.V.; Dzantiev, B.B. Method for Carrying out Highly Sensitive Immunochromatographic Analysis with Cascade Multi-Stage Signal Amplification. Russian Federation Patent Application No. 2015153293, 14 December 2015.
32. Hackett, J.D.; Anderson, D.M.; Erdner, D.L.; Bhattacharya, D. Dinoflagellates: A remarkable evolutionary experiment. *Am. J. Bot.* **2004**, *91*, 1523–1534. [CrossRef]

foods

MDPI

Article

Design and Characterization of a Novel Hapten and Preparation of Monoclonal Antibody for Detecting Atrazine

Lingyuan Xu [1], A.M. Abd El-Aty [2,3], Jae-Han Shim [4], Jong-Bang Eun [5], Xingmei Lei [1], Jing Zhao [1], Xiuyuan Zhang [1], Xueyan Cui [1], Yongxin She [1], Fen Jin [1], Lufei Zheng [1], Jing Wang [1], Maojun Jin [1,*] and Bruce D. Hammock [6]

1 Institute of Quality Standard and Testing Technology for Agro-Products, Chinese Academy of Agricultural Sciences, Beijing 100081, China; xulingyuan163@163.com (L.X.); leixingmei163@163.com (X.L.); zj13048453218@163.com (J.Z.); zhangxiuyuan2017@163.com (X.Z.); cxy18763802353@163.com (X.C.); 0891syx@163.com (Y.S.); jinfenbj@163.com (F.J.); zhenglufei@caas.cn (L.Z.); w_jing2001@126.com (J.W.)
2 Department of Pharmacology, Faculty of Veterinary Medicine, Cairo University, Giza 12211, Egypt; abdelaty44@hotmail.com
3 Department of Medical Pharmacology, Medical Faculty, Ataturk University, Erzurum 25240, Turkey
4 Natural Products Chemistry Laboratory, Biotechnology Research Institute, Chonnam National University, Yongbong-ro, Buk-gu, Gwangju 500-757, Korea; jhshim@jnu.ac.kr
5 Department of Food Science and Technology, Chonnam National University, Gwangju 500-757, Korea; jbeun@jnu.ac.kr
6 Department of Entomology & Nematology and the UC Davis Comprehensive Cancer Center, University of California, Davis, CA 95616, USA; bdhammock@ucdavis.edu
* Correspondence: jinmaojun@caas.cn; Tel.: +86-10-8210-6570

Abstract: This study provides the first design and synthetic protocol for preparing highly sensitive and specific atrazine (ATR) monoclonal antibodies (mAbs). In this work, a previously unreported hapten, 2-chloro-4-ethylamino-6-isopropylamino-1,3,5-triazine, was designed and synthesized, which maximally exposed the characteristic amino group ATR to an animal immune system to induce the expected antibody. The molecular weight of the ATR hapten was 259.69 Da, and its purity was 97.8%. The properties of the anti-ATR mAb were systematically characterized. One 9F5 mAb, which can detect ATR, was obtained with an IC_{50} value (the concentration of analyte that produced 50% inhibition of ATR) of 1.678 μg/L for ATR. The molecular weight for the purified 9F5 mAb was approximately 52 kDa for the heavy chain and 15 kDa for the light chain. The anti-ATR mAb prepared in this study was the IgG_1 type. The working range of the standard curve (IC_{20} (the concentration of analyte that produced 20% inhibition of ATR)_IC_{80} (the concentration of analyte that produced 80% inhibition of ATR)) was 0.384 to 11.565 μg/L. The prepared anti-ATR mAb had high specificity, sensitivity, and affinity with low cross-reactivity. The prepared anti-ATR mAb could provide the core raw material for establishing an ATR immunoassay.

Keywords: atrazine; hapten; monoclonal antibody; immunoassay

Citation: Xu, L.; Abd El-Aty, A.M.; Shim, J.-H.; Eun, J.-B.; Lei, X.; Zhao, J.; Zhang, X.; Cui, X.; She, Y.; Jin, F.; et al. Design and Characterization of a Novel Hapten and Preparation of Monoclonal Antibody for Detecting Atrazine. *Foods* **2022**, *11*, 1726. https://doi.org/10.3390/foods11121726

Academic Editor: Simon Haughey

Received: 10 May 2022
Accepted: 7 June 2022
Published: 13 June 2022

Publisher's Note: MDPI stays neutral with regard to jurisdictional claims in published maps and institutional affiliations.

Copyright: © 2022 by the authors. Licensee MDPI, Basel, Switzerland. This article is an open access article distributed under the terms and conditions of the Creative Commons Attribution (CC BY) license (https://creativecommons.org/licenses/by/4.0/).

1. Introduction

Atrazine (2-chloro-4-diethylamino-6-isopropylamino-1,3,5-triazine, ATR) is an extensively used selective systemic triazine herbicide [1]. It resists environmental degradation owing to a slow photolysis and hydrolysis rate [2]. It disrupts the photosynthetic (energy-producing) process in broadleaf weeds and annual grasses [3]. ATR has become the most frequently detected pesticide in surface water and groundwater [4]. It has a long residence time and a stable structure. Therefore, it remains active for several years, causing environmental pollution [5]. Because of its low toxicity, it allows affected organisms to survive long-term, inducing subacute damage. Large amounts of pesticide residues thus remain in crops, water, and soil. This can cause severe damage to living organisms and disrupt the ecological balance [6]. ATR has carcinogenicity, teratogenicity, and mutagenicity. It affects the reproductive, endocrine, central nervous, and immune systems [7–9]. Therefore, it is imperative to establish a rapid and highly sensitive qualitative and quantitative detection technique for ATR residues.

Many countries have established residue limits for ATR. The European Food Safety Authority (EFSA) has assessed the maximum residue limit (MRL) of ATR in cereals to be 0.1 mg/kg. The European Union requires that the mass concentration of a single pesticide in drinking water should not exceed 0.1 μg/L and that the total mass concentration of all pesticides should not exceed 0.5 μg/L. In Australia, the MRL for ATR in corn, sugarcane, and sorghum is 0.1 mg/kg. The US EPA has set the MRL for ATR in fat, meat, and meat byproducts to 0.02 mg/kg and the MRL for ATR in drinking water and corn to 0.1 μg/kg and 0.25 mg/kg, respectively. In China, the MRL for ATR in drinking water (GB 5749-2006) and surface water (GB 3838-2002) is 2 and 3 μg/L, respectively.

The conventional detection methods used for ATR residue analysis include instrumental and immunoassay methods. Chromatographic analysis methods include gas chromatography (GC) [10], gas chromatography–mass spectrometry (GC/MS) [11], high-performance liquid chromatography (HPLC) [12], liquid chromatography–mass spectrometry (HPLC/MS) [13], and ultra high performance liquid chromatography (UHPLC) [14]. These methods are highly accurate and sensitive. However, they require complex sample pretreatment, expensive instrumentation, and cumbersome operation, which do not allow for extensive testing. Immunological assays have the advantages of simplicity, rapidity, and accuracy, meeting the requirements of modern rapid detection techniques. Immunoassays are based on the principle of the specific recognition of antigens and antibodies. Antibodies serve as the basis for immunological detection methods. The preparation of antibodies with high specificity, sensitivity, and affinity has become the most important part of improvements in detection techniques.

This study designed and synthesized an ATR hapten to retain part of its active structure. The purity, molecular weight, and structure were identified. The prepared ATR hapten was pure and structurally accurate. The ATR complete antigen was prepared using the active ester method. Mouse monoclonal antibodies against ATR were obtained by immunizing mice using immunogen, cell fusion, and hybridoma screening. The anti-ATR monoclonal antibody prepared in this study was the IgG_1 type. The availability of monoclonal antibodies can lay the foundation for establishing immunoassays, enabling the rapid and high-throughput detection of ATR. The high specificity, sensitivity, and affinity of the murine monoclonal antibody against ATR could meet the requirements for the rapid, high sensitivity, and selectivity screening of ATR residues in agricultural products and environmental contamination in the field.

2. Materials and Reagents

Melamine, sodium bicarbonate, ethanol, and dichloromethane were procured from Beijing Chemical Factory Co. (Beijing, China) N,N-diisopropylethylamine (DIEA) and 3-aminobutyric acid were acquired from Beijing Evenhe Technology Co. (Beijing, China). The ATR standard was acquired from First Standard (Tianjin, China). N,N'-Dicyclohexylcarbodiimide (DMF) was supplied by Tianjin Seans Biochemical Technology Co. (Tianjin, China). 1-Ethyl-(3-dimethylaminopropyl) carboimide hydrochloride (EDC), N-hydroxysuccinimide (NHS), polyethylene glycol 2000(PEG-2000), cell freezing medium dimethyl sulfoxide (DMSO; serum-free), hypoxanthine, aminopterin, and thymidine (HAT), hypoxanthine and thymidine (HT) medium supplements, penicillin, streptomycin, L-glutamine, horse-radish-peroxidase-labeled goat anti-mouse IgG, complete and incomplete Freund's adjuvant, TMB (3,30,5,50-tetramethyl benzidine) substrate solution, 1-(3-dimethylaminopropyl)-3-ethylcarbodiimide hydrochloride (EDC), bovine serum albumin (BSA), ovalbumin (OVA), 50% (*w*/*v*) polyethylene glycol solution, and dimethyl sulfoxide (DMSO) were provided by Sigma–Aldrich (St. Louis, MO, USA). Cell culture medium (Dulbecco's modified Eagle's medium; DMEM) and fetal bovine serum (FBS) were obtained from Gibco BRL (Paisley, Scotland). GIBCO® Australian Premium FBS, GIBCO® DMEM basic (1X) basal culture medium, penicillin solution, and L-glutamine solution were picked up from Thermo Fisher Scientific(Waltham, MA, USA). Analytical-grade sodium bicarbonate ($NaHCO_3$), sodium carbonate (Na_2CO_3), sodium chloride (NaCl), sodium hydrogen phosphate dodecahydrate disodium hydrogen phosphate dodecahydrate ($Na_2HPO_4 \cdot 12H_2O$), potassium dihydrogen phosphate (KH_2PO_4), citric acid monohydrate ($C_6H_{10}O_8$), Tween-20, hydrogen peroxide (H_2O_2), and gelatin were purchased from Sinopharm Chemical Reagent Co. (Beijing, China). The mouse Sp2/0-Ag14 myeloma cell line was purchased from the Cell Resource Center of Peking Union Medical College (Beijing, China). Triazine standards (ATR, terbuthylazine, simetryn, propytryn, terbuthylazine, simazine; 99% purity) were secured from First Standard (Tianjin, China). Horseradish peroxidase (HRP) was purchased from Jackson Immunoresearch Laboratories Co. (West Grove, PA, USA). Cell culture plates (6-well™, 24-well™, and 96-well™) and a 96-MicroWell™ transparent plate were secured from Costar (Corning, USA). Cell pipettes (1, 5, and 10 mL) were purchased from Thermo Fisher Scientific (Thermo, Vantaa, Finland). Syringes (1 and 5 mL) were purchased from Shandong Zhu Pharmaceutical Co. (Shandong, China). The glassware used in the experiments was strictly cleaned and sterilized before use. The glassware, tips, and reagents were sterilized by autoclaving. The autoclave was set at 120 °C, and items were sterilized for 20 min. After sterilization, they were placed in an oven, allowed to dry, and then used.

3. Methods

3.1. Synthesis of ATR Hapten

The synthesis route of the hapten is shown in Figure 1a. Briefly, the ATR hapten was first obtained by synthesizing product SM_1 and then reacting SM_1 with 3-aminobutyric acid to obtain the desired hapten. The intermediate SM_1 was synthesized by dissolving melamine (0.92 g, 5.0 mmol) in acetonitrile (50 mL). The solution was cooled to 0 °C in an ice-water bath. The cooled solution was mixed with an aqueous solution of ethylamine (0.99 mL, 5.0 mmol) and diisopropylethylamine (DIEA) (2.65 mL, 15.0 mmol). The reaction mixture was stirred at 0 °C for 4 h and then dried by spin evaporation. The crude product obtained was separated by silica gel column chromatography (petroleum ether:ethyl acetate = 5:1) to obtain product SM_1 (4,6-dichloro-N-ethyl-1,3,5-triazine-2-amine), 780 mg (81%). SM_2 was synthesized by dissolving SM_1 (935 mg, 5 mmol) in 20 mL ethanol. Then, SM_2 (515 mg, 5 mmol) and DIPEA (1.94 g, 15 mmol) were mixed. The reaction solution was heated to 85 °C for 3 h. The complete reaction of SM_1 was monitored using thin-layer chromatography (TLC). The solution was then cooled to room temperature at 25 °C and spin-dried. Saturated aqueous sodium bicarbonate solution (20 mL) was added to the spin-dried sample and stirred for 20 min. The reaction was completed, and the products were extracted twice with 20 mL of dichloromethane. The aqueous phase was adjusted with 1 M HCl (pH = 4). The white precipitate obtained from the reaction was filtered. After drying, 850 mg of ATR hapten was obtained. The recovery rate from this reaction was 62.5%. Finally, the hapten was characterized for purity, molecular weight, and structure using high-performance liquid chromatography (HPLC), high-resolution mass spectrometry (HRMS), nuclear magnetic resonance hydrogen spectroscopy (^1H NMR), and carbon spectroscopy (^{13}C NMR).

3.2. Preparation of Immunogen and Coating Antigen of ATR

A small-molecule hapten is only antigenic but not immunogenic. The immunogenicity of a small molecule hapten is achieved when it is linked to a large molecule (usually a protein). BSA has stable physical and chemical properties and many free amino groups and is cheap and easy to obtain. It still maintains great solubility under different pH values and ionic strengths and contains some organic solvents. It is often used as the carrier of immunogens. OVA can be used as a carrier protein for antibody screening and immunoassays because of its weak immunogenicity compared with other proteins. In this study, BSA and OVA were coupled with ATR hapten to prepare immunogen and coating antigen. The link between the hapten and carrier protein mainly occurs by coupling carboxyl and amino groups and other active wave groups. To obtain an ATR complete antigen with immunogenicity, the carboxyl group of the hapten was activated using the activating ester method and coupled to the carrier protein. The structural formulae for the activation of the hapten and the protein coupling reaction are shown in Figure 1b. The molar ratio of the hapten to the immunogen was 60:1, and the coating antigen was 50:1. The specific protocol was to add 3.15 mg (0.012 mmol) 3-(4-chloro-6-(ethylamino)-1,3,5-triazine-2-ylamino) butanoic acid, 2.79 mg (0.024 mmol) NHS, and 4.60 mg (0.024 mmol) EDC in 0.5 mL DMF. The solution was stirred magnetically overnight (10 h) in a refrigerator at 4 °C. After the reaction was completed, an activation solution containing the ATR hapten was obtained. This activation solution can be directly used for subsequent coupling with carrier proteins. After that, 20 mg of BSA (or 10 mg of OVA) was dissolved in 0.01 M PBS buffer (pH = 7.4, 10 mg/mL). Activated hapten was added to the BSA and OVA solutions drop by drop. The reactions were conducted under magnetic stirring at room temperature for 4 h. The resulting reaction solution was then dialyzed six times with 4 L of 0.01 mol/L PBS (pH 7.4). The dialysis removed unreacted hapten or other small molecules. After dialysis was completed, the antigen was dispensed at 1 mg/mL, snap-frozen in liquid nitrogen, and stored at −20 °C.

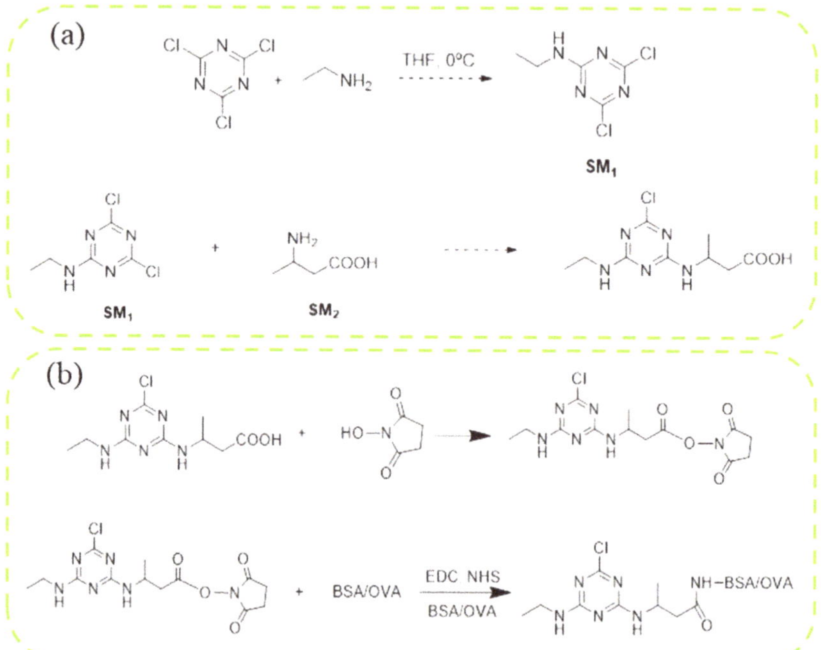

Figure 1. Synthetic routes of hapten (**a**) and immunogen and coating antigen (**b**) of ATR.

Immunogenicity was characterized by matrix-assisted laser desorption/ionization time-of-flight mass spectrometry (MALDI-TOF-MS). MALDI-TOF-MS measured the conjugation ratio of the coating antigen and immunogen. This method has the advantages of rapid and straightforward sample preparation and high sensitivity. The matrix auxiliary solution used for the measurement was configured as follows: the mixture (acetonitrile:water = 70:30) contained 0.001% trifluoroacetic acid and erucic acid (15 mg/mL). The wavelength of the Yag excitation light source was 355 nm. The m/z acquisition range was between 10 and 100 kDa. Each sample was mixed with the matrix auxiliary solution. After mixing, spot sample 1 was placed on the sample target of 1 μL. After natural drying, the sample was placed under a mass spectrometer for Yag laser scanning. The conjugation ratio was calculated as follows:

$$\text{Conjugation ratio} = (M_p - M_{std})/M_h.$$

where "M_p" represents the ATR antigen conjugates, "M_{std}" represents the BSA/OVA standard, and "M_h" represents the ATR hapten.

4. Production and Characteristics of Anti-ATR mAb

4.1. Buffers and Solutions

(1) Coating buffer solution (CBS, pH 9.6, 0.05 mol/L): 2.93 g $NaHCO_3$ and 1.5 g Na_2CO_3 were weighed and dissolved with Milli-Q water, and the volume was adjusted to 1 L. (2) Phosphate buffer solution (PBS, pH 7.4, 0.01 mol/L): 0.2 g KH_2PO_4, 2.96 g $Na_2HPO_4 \bullet 12H_2O$, and 8.0 g NaCl were weighed and dissolved in Milli-Q water, and the volume was adjusted to 1 L. (3) Washing buffer solution (PBST): PBS containing 0.1% Tween-20. (4) Sample dilution buffer solution (PBSTG): PBST containing 0.1% gelatin. (5) TMB chromogenic solution: three solutions are currently used and configured. The formula of each enzyme label plate was as follows: 11 mL matrix buffer (46.04 g potassium dihydrogen citrate hydrate and 0.10 g potassium sorbate dissolved to 1 L and stored at room temperature), 200 μL TMB stock solution (375 mg, 3,3,5,5′-tetramethylbenzidine dissolved in 30 mL dimethyl sulfoxide) and 101 μL 1% H_2O_2 (1 mL 30% H_2O_2 added to 29 mL Milli-Q water). (6) Termination solution (1 mol/L HCl): 44 mL of 98% hydrochloric acid was measured, 440 mL of deionized water was slowly added along the beaker wall, and the mixture was stirred while adding.

4.2. Immunization

Animal experiments were approved by the Experimental Animal Welfare and Ethical Committee of Institute of Quality Standards and Testing Technology for Agro-Products, Chinese Academy of Agricultural Sciences (IQSTAP-2021-05). Animal experiments were conducted in strict accordance with Chinese laws and guidelines. Six female Balb/c mice (7 weeks old) were immunized with immunogen (1 mL ATR-BSA (1 mg/mL, molar ratio of 19:1) + 1 mL Freund's adjuvant (complete/incomplete)). The immunogen was mixed with Freund's complete adjuvant for the initial immunization. Subsequent immunizations were performed by fully emulsifying the immunogen with an equal volume of Freund's incomplete adjuvant. The immunization strategy for the mice is shown in Table 1. At 3–5 d after the third injection, the immunized mice were eye-bled, and the sera were tested for anti-ATR antibody titer and ATR recognition properties using ic-ELISA. The ic-ELISA protocol, buffers, and solutions were similar to those described previously [15]. The reaction was terminated by adding 50 μL of 1 M HCl per well. Optical density (OD) values at 450 nm were measured with an Infinite M200 PRO microplate reader (TECAN, Männedorf, Switzerland). The specificity of the developed anti-ATR serum was assessed using ic-ELISA. The inhibition rate was calculated according to the following equation:

$$IR\ (\%) = (1 - B/B_0) \times 100\%$$

where "IR" represents the inhibition rate, B represents the OD_{450nm} value of the inhibition well, and B_0 represents the OD_{450nm} value of the control wells.

Table 1. Immunization strategy in mice.

Number of Immunizations	Immunization Cycle (Days)	Total Volume of Immunization (mL)	Immunization Dose (mg)
First	1	0.2 [a]	0.1
Second	15	0.2 [b]	0.1
Third	29	0.2 [b]	0.1
Fourth	43	0.1	0.05

* [a]—complete Freund's adjuvant, [b]—incomplete Freund's adjuvant.

4.3. Ic-ELISA Procedure

Ic-ELISA was applied to screen the serum of mice with the best performance fusion, detect the positive cell supernatant, and establish the standard curve. The operation steps are as follows: (1) Add the coated antigen to the microplate, and incubate each well with 100 μL at 37 °C for 30 min. Then, wash the plate with PBST 3 times. (2) Dilute the standard sample and antibody with PBSTG to the required concentration, add standard samples of different concentrations and 50 μL 9F5mAb in the microplate, and incubate at 37 °C for 30 min; wash the plate with PBST 3 times. (3) Add 100 μL IgG HRP diluted with PBSTG to each well; wash the plate with PBST 3 times. (4)Add 100 μL chromogenic solution to each well, avoid light at 25 °C for 15 min, and then add 50 μL hydrochloric acid solution to each well to terminate the reaction. The absorbance value was measured at $OD450_{nm}$. In the whole process, 200 μL PBST was added to each well when washing the enzyme label plate.

4.4. *Production of Anti-ATR mAb*

4.4.1. Preparation of Sp2/0 and Feeder Cells

One week before cell fusion, mice of the same strain used for culture and immunization were resuscitated and expanded. The specific operation steps of Sp2/0 (obtained from National Infrastructure of Cell Line Resource, Beijing, China) for myeloma cells are as follows: quickly transfer the cells from the liquid nitrogen tank to the 37 °C water bath until the solid is completely melted, and confirm that the cap of the cryopreservation tube has been tightened to prevent the cap from submerging into the water to prevent ice crystals from damaging the cell membrane. Preheat the DMEM culture solution to 37 °C in advance, slowly transfer the melted cell suspension to a 15 mL centrifuge tube containing 10–12 mL DMEM, blow it and mix it evenly, centrifuge at 1000 rpm/min for 10 min, and discard the supernatant. The cells were rung and dispersed at the bottom of the centrifuge tube on the super clean table. Then, 5 mL of 20% complete medium was slowly added to the cells to avoid too fast dropping acceleration and too large a change in cell osmotic pressure, resulting in cell wall rupture. After the resuspended cells were blown with a 5 mL pipette, they were transferred to a six-well plate with 2.5 mL per well and placed in a constant temperature incubator at 37 °C and 5% CO_2. When the cells grew to 80% of the area at the bottom of the six-well plate, that is, in the logarithmic growth stage, they continued to be subcultured with 20% complete culture medium. The culture was expanded for two to three generations, and bone marrow cells with good growth state, uniform size, and good shading were selected for the subsequent fusion experiment of splenocytes and tumor cells.

A healthy female mouse over 10 weeks old was selected, the eyeball was removed, bleeding to death occurred, and the mouse was soaked in alcohol for 5 min. Fix the mouse with a needle on the anatomical table, with the abdomen upward, cut the epidermis with scissors, clamp the peritoneum with sterile tweezers, inject liquid with a disposable needle tube, gently squeeze the mouse's abdomen with tweezers or shake its legs, draw out the liquid with a needle tube and repeat several times. The liquid was added to the extracted liquid, the supernatant was centrifuged and discarded, and the cells were resuspended in culture medium and placed in 96-well culture plates at 100 μL per well.

4.4.2. Cell Fusion and Screening

The spleen cells collected from the mice were fused with the SP2/0 cell line using PEG-2000 at a ratio of 10:1 spleen:myeloma cells. The serum titer reached 6.4×10^4 at an OD450 nm value higher than 1.0. This indicated good immunogenicity. The mice with the best inhibitory effect were selected to enhance immunity. Intraperitoneal injections of 0.05 mg immunogen (2 mg/mL, 0.025 mL) were conducted for the fusion. Three days after the last immunization booster, mouse spleen cells were separated and fused with PEG-2000 pretreated Sp2/0 myeloma cells to prepare hybridomas according to previously described procedures [16–18]. The cell culture medium formulations are listed in Table 2. The plates were incubated at 37 °C in a CO_2 incubator (5% CO_2 in air). The selective growth of the hybrid cells occurred in DMEM supplemented with 2% HAT. The fused cells were cultured in 2% HAT medium for 7 day. Several hybrid cells were screened seven days after fusion by testing the supernatant using ic-ELISA to determine the binding ability. Positive hybridomas were cloned by limiting dilution, and clones were further selected by ic-ELISA [19–21]. The anti-ATR mAb clone, designated 9F5, which had a high antibody titer and good sensitivity in the culture supernatant, was expanded in mice to produce mAb in ascites. The expanded cultured monoclonal cell line was injected into the abdominal cavity of mice treated with paraffin 1 week in advance to prepare the ascites antibody. Ic-ELISA was used to detect the efficacy and inhibition rate of mouse ascites. The antibodies were purified by ammonium sulfate precipitation [22,23]. The best antigen–antibody concentration was selected through checkerboard ic-ELISA and then established with the best combination. The standard curve was established to determine the sensitivity and detection range of the method. The purified antibody was dissolved in 0.01 M PBS and placed into a dialysis bag. It was dialyzed with 0.01 M PBS four times and PB two times, and the dialysate was changed every four hours. The dialyzed antibody protein was packed in a 1.5 mL centrifuge tube, frozen in liquid nitrogen, and vacuum lyophilized. The 9F5 mAb dissolved the lyophilized antibody powder with 0.01 M PBS and stored at −20 °C. Then, 50% glycerol was added and prepared at 1 mg/mL. This was used to detect the properties of monoclonal antibodies, including titer, specificity, and type. The assay cross-reactivity with ATR 9F5 mAb was determined using ic-ELISA. OriginPro 8.5 (OriginLab Corporation, Northampton, MA, USA) for Windows was used for data analysis. The OD_{450nm} values were plotted against the analyte concentration on a logarithmic scale, and the generated sigmoidal curve was mathematically fitted to a four-parameter logistic equation.

Table 2. Cell culture medium formulations.

Name of Culture Medium	Formulation Volume Ratio (%)					
	DMEM Basal Culture Solution	L-Glutamine Solution	Penicillin Solution	50 × HAT Recovery Solution	DMSO	Fetal Bovine Serum
Complete culture solution	78	1	1	–	–	20
HAT complete Culture solution	76	1	1	2	–	20
Cell lyophilization solution	69	1	1	–	9	20

*–: no.

5. Results and Discussion

5.1. Identification Results of Hapten, Immunogen, and Coating Antigen of ATR

5.1.1. Identification of Atrazine Hapten

The identification results of the ATR hapten are shown in Figure S1. The theoretical precise molecular weight $[M + {}^{+}H]$ of the ATR hapten $C_9H_{14}C_lN_5O_2$ was 259.69. According to the HPLC and HRMS results, the purity of the ATR hapten was 97.8%, as revealed by the peak area ratio. The peak positions indicated that the correct ATR hapten molecule was synthesized.

^{1}H NMR (400 M*Hz*, DMSO-d6) δ12.16 (s, 1H), 7.86–7.61 (m, 2H), 4.27 (h, J = 6.8 *Hz*, 1H), 3.24 (dt, J = 14.1, 6.9 *Hz*, 2H), 2.60–2.51 (m, 0H), 2.32 (dt, J = 15.8, 6.4 *Hz*, 1H), 1.18–1.02 (m, 6H).

^{13}C NMR (101 M*Hz*, DMSO) δ 172.37, 167.53, 165.07, 164.81, 164.54, 43.39, 43.12, 40.27, 40.12, 39.91, 39.70, 39.49, 39.28, 39.08, 38.87, 35.01, 34.84, 20.26, 20.00, 19.81, 14.59, 14.36, 14.23.

5.1.2. Identification of Immunogen and Coating Antigen of ATR

The hapten and BSA feed ratios were 40:1, 50:1, and 60:1, respectively. The hapten and OVA feed ratios were 30:1, 40:1, and 50:1, respectively. The single charge ion peaks of BSA and OVA standards were 67132.709 and 44587.269, respectively. The molecular weight of ATR was 259.69. The average number of ATR hapten molecules coupled on each BSA and OVA by the active ester method can be calculated from the formula. The

MALDI-TOF-MS results of the immunogen and coated antigen are shown in Figure S2. The coupling results of coated antigen and immunogen are listed in Table 3. According to the data, the immunogen selected by immunized mice was 60:1. The coated antigen ratio of ic-ELISA was 50:1.

Table 3. Identification of the immunogen and coated antigen of ATR.

(A) Identification results for the immunogen.				
Hapten to Carrier Protein Dosing Ratio	**Molecular Weight of Hapten-BSA (Da)**	**Mass Change Δm (Da)**	**Hapten to Carrier Protein Ratio**	**Hapten Number/BSA**
40:1	71,131.224	3998.515	14.46	14
50:1	71,413.430	4280.721	16.61	16
60:1	72,205.234	5072.525	19.05	19
(B) Identification results of coated antigen.				
Hapten and Carrier Protein Dosing Ratio	**Molecular Weight of Hapten-OVA (Da)**	**Mass Change Δm (Da)**	**Hapten to Carrier Protein Ratio**	**Hapten Number/OVA**
30:1	46,093.340	1415.697	5.45	5
40:1	46,113.244	1473.982	5.68	5
50:1	46,347.978	1727.179	6.65	6

5.2. Characterization and Cross-Reactivity of Anti-ATR mAb

5.2.1. Characterization of Anti-ATR 9F5 mAb

The serum potency and inhibition rate of ATR mice are shown in Table S1 in Supplementary Materials. The serum titer and inhibition rate of the fusion mice are shown in Table S2. The titer and inhibition rate of mouse ascites before purification are shown in Table S3. The optimal antibody and coating antigen concentrations for the inhibition of 100 ng/mL ATR were determined using the ic-ELISA method. The determined optimal concentrations of the antigens and antibodies are shown in Table S4. Both the coated antigen and antibody were diluted to 2×10^3, 4×10^3, 8×10^3, 1.6×10^4, 3.2×10^4, 6.4×10^4, 1.28×10^5, and 2.56×10^5. They were used in subsequent experiments. The ATR standard was diluted to 100, 50, 25, 12.5, 6.25, 3.125, 1.5625, and 0 ng/mL in 0.01 M PBS for use in subsequent experiments. IC_{50} values from standard curves characterized the sensitivity of ic-ELISA under optimized conditions. The curve of ic-ELISA is shown in Figure 2a. Each value was the average of three independent repetitions. There was a good linear relationship between the inhibition rate of the antibody and the concentration of ATR. The IC_{50} of the anti-ATR 9F5 mAb was 1.678 μg/L, and the working range of the standard curve (IC_{20}–IC_{80}) was 0.384 to 11.565 μg/L.

According to the manufacturer's instructions, the molecular weight of the antibody was determined using an SDS–PAGE kit (Solarbio®, Beijing, China). The prepared separation and concentrated gel concentrations were 10% and 5%, respectively. The purified 9F5 monoclonal antibody showed two bands with molecular weights of approximately 52 and 15 kDa, that is, the molecular weights of the heavy and light chains of the mAb. Other miscellaneous protein bands were smaller with a higher antibody purity, indicating that these could be used to establish an ATR immunoassay method. The SDS–PAGE results are shown in Figure 2b. The immunoglobulin isotype was determined with a mouse antibody isotyping kit (Sigma, St. Louis, MO, USA). Three replicate groups were used in the experiment. According to Table 4, the type of 9F5 ATR monoclonal antibody is IgG_1.

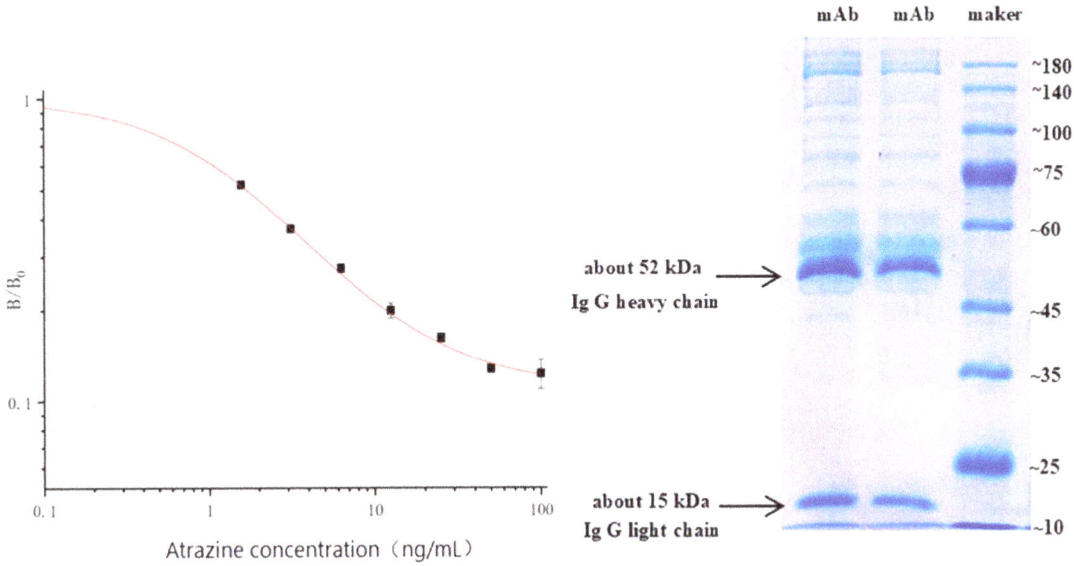

(a) (b)

Figure 2. Ic-ELISA curve of the 9F5 mAb (**a**) and the SDS–PAGE results (**b**).

Table 4. ATR 9F5 mAb type.

Antibody Typing	Number of Times	IgG_{2a}	IgG_{2b}	IgG_3	IgM	IgA	IgG_1
	1	0.9694	0.9774	0.9715	0.9704	0.9639	1.4213
OD_{450nm} value	2	1.014	1.0106	0.9461	0.9825	0.984	1.4076
	3	0.9891	0.9609	0.9363	1.0191	1.0386	1.4222

5.2.2. Cross-Reactivities of Anti-ATR mAb

The structure and cross-reaction rate of seven triazine herbicides are shown in Table 5. Although the cross-reaction rate between ATR and propazine was 41.50%, the cross-reactivities with other structural analogs were low. The existence of cross-reactions with other triazine herbicides is unavoidable because the chemical structures of triazine herbicides are very similar. This shows that the prepared 9F5 anti-ATR mAb had good specificity for ATR.

Table 5. Structures of triazine herbicides and their cross-reactivity rates.

Analytes	Chemical Structure	IC_{50} (ng/mL)	Cross-Reactivity
Atrazine	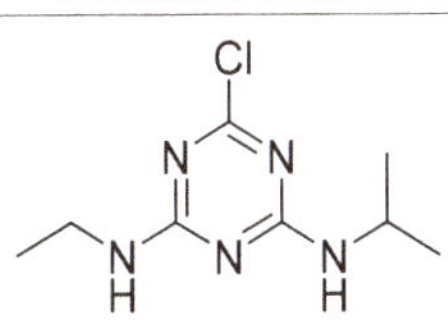	1.678	100.00%

Table 5. *Cont.*

Analytes	Chemical Structure	IC_{50} (ng/mL)	Cross-Reactivity
Propazine	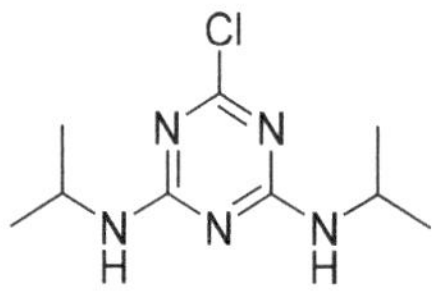	4.04	41.50%
Atrazine hapten	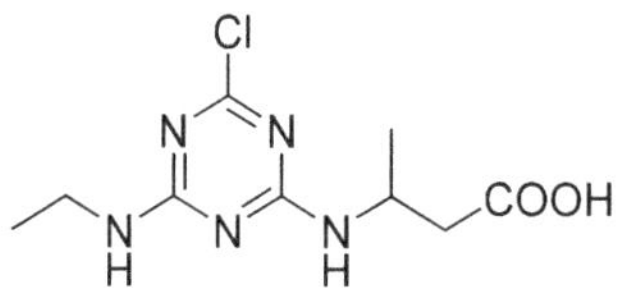	5.47	30.69%
Ametryn	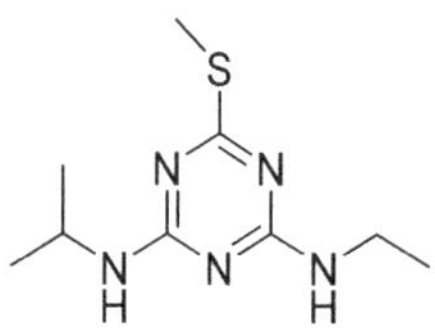	8.90	18.86%
Terbuthylazine	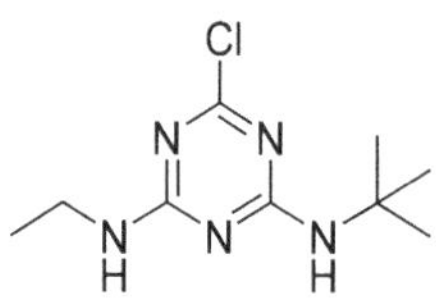	12.37	13.57%
Simazine	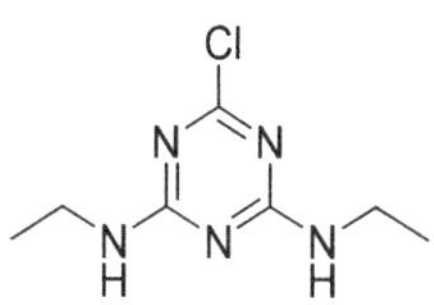	12.27	13.68%
Prometryn	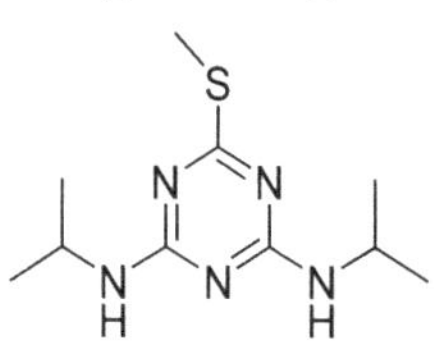	32.41	5.18%
Simetryn	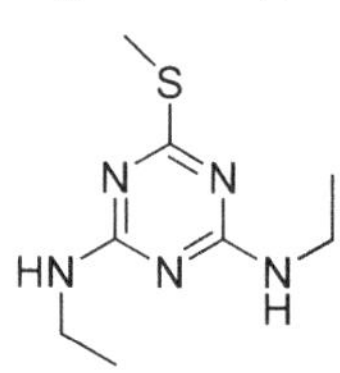	37.93	4.42%
Terbumeton	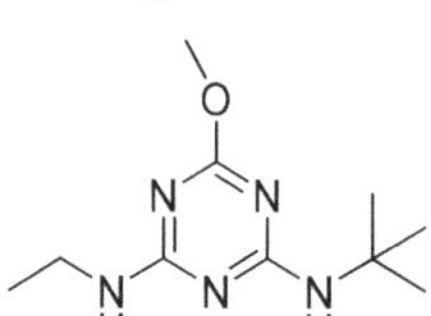	82.81	2.03%

Note: The concentrations of triazine pesticides were 100, 50, 25, 12.5, 6.25, 3.125, 1.5625, and 0 ng/mL.

6. Conclusions

In this study, a novel 2-chloro-4-ethylamino-6-isopropylamino-1,3,5-triazine (ATR hapten) was designed and synthesized. The ic-ELISA method was initially developed, however, the condition parameters were not optimized. Furthermore, the preliminarily developed ic-ELISA was not applied for the determination of actual samples. It should be noted that the prepared anti-ATR mAb can act as a raw material for establishing an ATR immunoassay.

The hapten retained the chlorine substituent at R_1, highlighting the antigenic determinant. Modifying the isopropylamine substituent at R_3 and adding the connecting arm and active groups to the isopropylamine substituent can effectively reduce the cross-reaction rate of the antibody to compounds containing isopropylamine substituents to improve the specific recognition of triazine pesticides by the atrazine antibody.

The molecular weight of the ATR hapten was 259.69 Da, and its purity was 97.8%. The characteristics of the anti-ATR mAb were systematically investigated. The mAb 9F5, which can specifically detect ATR, was obtained, and its IC_{50} value was 1.678 μg/L. The anti-ATZ mAb prepared in this study is IgG_1. The working range of the standard curve (IC_{20}–IC_{80}) was 0.384 to 11.565 μg/L. The molecular weights of the heavy and light chains of the purified 9F5 mAb were approximately 52 and 15 kDa, respectively. The prepared anti-ATR monoclonal antibody had high specificity, sensitivity, and affinity and low cross-reactivity.

The general structural formula of triazine herbicides is shown in Figure 3. The molecular formula of triazine herbicides that cross-react with mAb 9F5 is shown in Table 6. The previously reported atrazine hapten selectively introduces a carbon linkage arm and an active group (-CH_2)nCOOH) at the R_1 or R_2 position. In this study, the cross-reaction with five other triazine herbicides with similar structures was determined. It can be seen that the antibody has a high cross-reactivity with propazine and ametryn, which are 41.50% and 18.86%, respectively. The reason for this is that their structures are very similar. The two compounds have an isopropylamine substituent at the R_3 position. The designed hapten structure and cross-reaction rate are compared with previous reports, as shown in Table 7 [24–27]. The atrazine hapten prepared in this paper also has an ethylamine substituent at the R_3 position. Therefore, the ethylamine substituent at the R_2 position is a very important antigenic determinant of hapten. Some compounds without isopropylamine substituents and with methylthio or ethoxyl groups instead of Cl substituents at the R_1 position showed low cross-reaction rates. In summary, we can conclude that the isopropylamine substituent at R_3, Cl substituent at R_1, and ethylamine at R_2 also play a key role in the antigen–antibody recognition reaction.

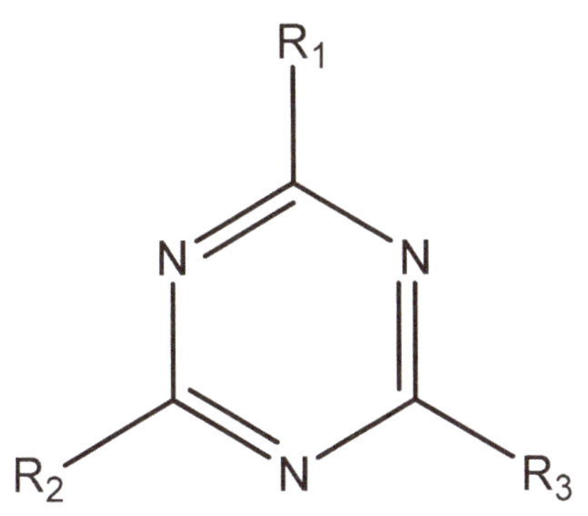

Figure 3. General structural formula of triazine herbicides.

Table 6. Molecular formula of atrazine and triazine herbicides with cross-reaction.

Name	R_1	R_2	R_3
Atrazine	Cl	$NHCH_2CH_3$	$NHCH(CH_3)_2$
Atrazine hapten	Cl	$NHCH_2CH_3$	$NHCH(CH_3)_2COOH$
Prometryne	SCH_3	$NHCH(CH_3)_2$	$NHCH(CH_3)_2$
Economicpoison	OCH_3	$NHCH_2CH_3$	$NHC(CH_3)_3$
Simazine	Cl	$NHCH_2CH_3$	$NHCH_2CH_3$
Terbuthylazine	SCH_3	$NHCH_2CH_3$	$NHC(CH_3)_3$
Simetryn	SCH_3	$NHCH_2CH_3$	$NHCH_2CH_3$

Table 7. Comparison with previously reported atrazine hapten structures and cross-reaction rates.

Hapten Name	Hapten Structure	CR(%)	Reference
4-chloro-6- (isopropyl amino)-l,3,5-triazine-2-(6-aminohexanecarboxylic acid)	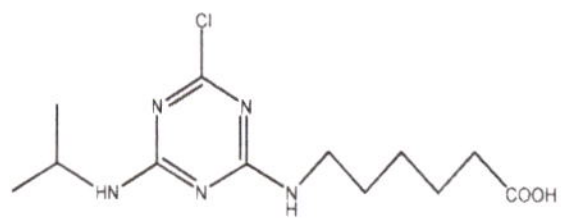	propazine (136%) cyanazine (28%) terbuthylazine (26%)	[24]
		AT-1-M8, AT-2-M1, and AT-3-M15 simazine (53–78%) AT-1-M3 MAb simazine (23%)	[25]
2-mercaptopropionic acid-4-ethylamino-6-isopropylamino-1,3,5-triazine	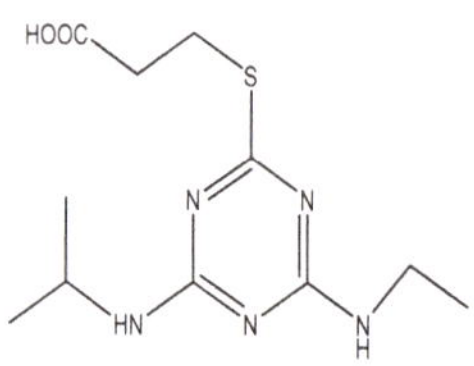	Simazine (19%) Melamine (0.043%) Chlorpyrifos (0.00012%) Monocrotophos (7.29×10^{-7}%) Parathion (0.0079%)	[26]
4-((4-(isopropylamino)-6-methyl-1,3,5-triazin-2-yl)amino)butanoic acid	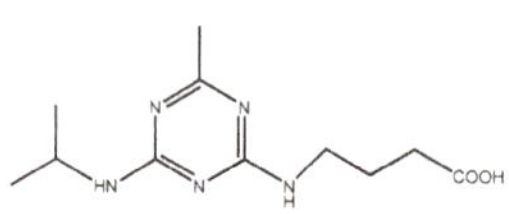	Simazine (2.5%) Propazine (90%) terbutylazine (5.0%) trietazine (11%) desmetryne (0.5%) ametryne (5.0%) atratone (6.4%) hydroxyatrazine (2.1%) hydroxysimazine (0.3%) hydroxypropazine (8.2%) hydroxyterbutylazine (5.6%) hydroxydesmetryne (0.1%)	[27]
2-chloro-4-ethylamino-6-isopropylamino-1,3,5-triazine	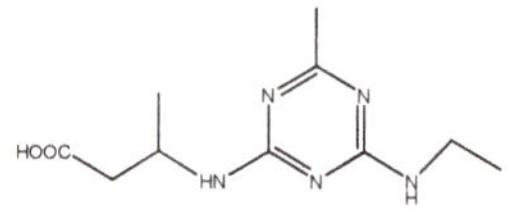	Propazine (41.50%) Atrazine hapten (30.69%) Ametryn (18.86%) Terbuthylazine (13.57%) Simazine (13.68%) Prometryn (5.18%) Simetryn (4.42%) Terbumeton (2.03%)	In this work

Supplementary Materials: The following supporting information can be downloaded at: https://www.mdpi.com/article/10.3390/foods11121726/s1, Table S1: Serum potency and inhibition rate of ATZ in mice; Table S2: Serum titer and inhibition rate of mouse for fusion experiment; Table S3: Titer and inhibition rate of mouse ascites before purification; Table S4: Determination of the optimal concentrations of antigen and antibody; Figure S1: Identification of ATR hapten using HPLC (a), HRMS (b), 1H NMR (c), and 13C NMR (d) analysis; Figure S2: MALDI-TOF-MS results of the immunogen (a) and coated antigen (b).

Author Contributions: Conceptualization, L.X. and M.J.; A.M.A., validation; formal analysis, L.X.; writing—review and editing; methodology, L.X.; software, L.X.; validation, L.X., X.L., X.C. and J.Z.; formal analysis, L.X.; investigation, L.X. and X.Z.; resources, L.X.; data curation, J.-H.S. and J.-B.E.; writing—original draft preparation, L.X.; writing—review and editing, L.X.; visualization, L.X.; supervision, Y.S., F.J., L.Z.; project administration, M.J.; funding acquisition, M.J., J.W. and B.D.H. All authors have read and agreed to the published version of the manuscript.

Funding: This work was supported by the Central Public Interest Scientific Institution Basal Research Fund for the Chinese Academy of Agricultural Sciences (Grant No.: Y2021PT05), (NIEHS) RIVER AWARD (Grant No.: R35ES030443), (NIEHS) Superfund Research Program (Grant No.: P42 ES004699), and Ningbo Innovation Project for Agro-Products Quality and Safety (Grant No.: 2019CXGC007).

Institutional Review Board Statement: The animal study protocol was approved by Experimental Animal Welfare and Ethical Committee of Institute of Quality Standards and Testing Technology for Agro-Products, Chinese Academy of Agricultural Sciences (IQSTAP-2021-05).

Informed Consent Statement: The study did not involve.

Data Availability Statement: The study did not report any data.

Conflicts of Interest: The authors declare no conflict of interest.

References

1. Jaeger, L.L.; Jones, A.D.; Hammock, B.D. Development of an enzyme-linked immunosorbent assay for atrazine mercapturic acid in human urine. *Chem. Res. Toxicol.* **1998**, *11*, 342–352. [CrossRef] [PubMed]
2. Ying, G.G.; Kookana, R.S.; Mallavarpu, M. Release behavior of triazine residues in stabilized contaminated soils. *Environ. Pollut.* **2005**, *134*, 71–77. [CrossRef] [PubMed]
3. Jablonowski, N.D.; Krutz, J.L.; Martinazzo, R.; Zajkoska, P.; Hamacher, G.; Borchard, N.; Burauel, P. Transfer of Atrazine Degradation Capability To Mineralize Aged 14 C-Labeled Atrazine Residues in Soils. *J. Agric. Food Chem.* **2013**, *61*, 6161–6166. [CrossRef]
4. Jin, R.; Ke, J. Atrazine and its degradation products in surface and ground waters in Zhangjiakou District, China. *Chin. Sci. Bull.* **2002**, *47*, 1612–1616.
5. Czaplicka, M.; Barchanska, H.; Jaworek, K.; Kaczmarczyk, B. The interaction between atrazine and the mineral horizon of soil: A spectroscopic study. *J. Soil Sediments* **2018**, *18*, 827–834. [CrossRef]
6. Hayes, T.B.; Collins, A.; Lee, M.; Mendoza, M.; Noriega, N.; Stuart, A.A.; Vonk, A. Hermaphroditic, demasculinized frogs after exposure to the herbicide atrazine at low ecologically relevant doses. *Proc. Natl. Acad. Sci. USA* **2002**, *99*, 5476–5480. [CrossRef]
7. Kornilovskaya, I.N.; Gorelaya, M.V.; Usenko, V.S.; Gerbilsky, L.V.; Berezin, V.A. Histological studies of atrazine toxicity on the thyroudgland in rats. *Biomed. Environ. Sci.* **1996**, *9*, 60–66.
8. Podda, M.V.; Deriu, F.; Solinas, A.; Demontis, M.P.; Varoni, M.V.; Spissu, A.; Anania, V.; Tolu, E. Effect of atrazine administration on spontaneous and evoked cerebellar activity in the rat. *Pharmacol. Res.* **1997**, *36*, 199–202. [CrossRef]
9. Thomas, S.J.; Willem, S.; Giesy, J.P.; Martin, V.D.B. 2-chloro-s-triazine herbicides induce aromatase (CYP 19) activity in H295R human adrenocortical carcinoma cells: A novel mechanism for estrogenicity? *Toxicol. Sci.* **2000**, *54*, 121–127.
10. Jia, L.C.; Su, M.; Wu, X.Q.; Sun, H.W. Rapid selective accelerated solvent extraction and simultaneous determination of herbicide atrazine and its metabolites in fruit by ultra high-performance liquid chromatographyy. *J. Sep. Sci.* **2016**, *39*, 4512–4519. [CrossRef]
11. Djozan, D.; Ebrahimi, B. Preparation of new solid phase micro extraction fiber on the basis of atrazine-molecular imprinted polymer: Application for GC and GC/MS screening of triazine herbicides in water, rice and onion. *Anal. Chim. Acta* **2008**, *616*, 152–159. [CrossRef] [PubMed]
12. Wu, Q.; Li, Z.; Wu, C.; Wang, C.; Wang, Z. Application of ultrasound-assisted emulsification microextraction for the determination of triazine herbicides in soil samples by high-performance liquid chromatography. *Microchim. Acta* **2010**, *170*, 59–65. [CrossRef]
13. Katsumata, H.; Kojima, H.; Kaneco, S.; Suzuki, T.; Ohta, K. Preconcentration of atrazine and simazine with multiwalled carbon nanotubes as solid-phase extraction disk. *Microchem. J.* **2010**, *96*, 348–351. [CrossRef]
14. Panuwet, P.; Nguyen, J.V.; Kuklenyik, P.; Udunka, S.O.; Needham, L.L.; Barr, D.B. Quantification of atrazine and its metabolites in urine by on-line solid-phase extraction-high-performance liquid chromatography-tandem mass spectrometry. *Anal. Bioanal. Chem.* **2008**, *391*, 1931–1939. [CrossRef]
15. Zhao, J.; Li, G.; Wang, B.M.; Liu, W.; Nan, T.G.; Zhai, Z.X.; Zhao, H.L.; Li, Q.X. Development of a monoclonal antibody-based enzyme-linked immunosorbent assay for the analysis of glycyrrhizic acid. *Anal. Bioanal. Chem.* **2006**, *386*, 1735–1740. [CrossRef]
16. Mi, J.; Dong, X.; Zhang, X.; Li, C.; Wang, J.; Mujtaba, M.G.; Zhang, S.; Wen, K.; Yu, X.; Wang, Z. Novel hapten design, antibody recognition mechanism study, and a highly sensitive immunoassay for diethylstilbestrol in shrimp. *Anal. Bioanal. Chem.* **2019**, *411*, 5255–5265. [CrossRef]

17. Li, H.; Ma, S.; Zhang, X.; Li, C.; Dong, B.; Mujtaba, M.G.; Wei, Y.; Liang, X.; Yu, X.; Wen, K.; et al. Generic hapten synthesis, broad-specificity monoclonal antibodies preparation, and ultrasensitive ELISA for five antibacterial synergists in chicken and milk. *J. Agric. Food Chem.* **2018**, *66*, 11170–11179. [CrossRef]
18. Köhler, G.; Milstein, C. Continuous cultures of fused cells secreting antibody of predefined specificity. *Nature* **1975**, *256*, 495–497. [CrossRef]
19. Holzlöhner, P.; Hanack, K. Generation of murine monoclonal antibodies by hybridoma technology. *JoVE* **2017**, *119*, e54832. [CrossRef]
20. Zaroff, S.; Tan, G. Hybridoma technology: The preferred method for monoclonal antibody generation for in vivo application. *Biotechniques* **2019**, *67*, 90–92. [CrossRef]
21. Yagami, H.; Kato, H.; Tsumoto, K.; Tomita, M. Monoclonal antibodies based on hybridoma technology. *Pharm. Pat. Anal.* **2013**, *2*, 249–263. [CrossRef] [PubMed]
22. Carroll, S.; Al-Rubeai, M. The selection of high-producing cell lines using flow cytometry and cell sorting. *Expert Opin. Biol. Ther.* **2004**, *4*, 1821–1829. [CrossRef] [PubMed]
23. Browne, S.M.; Al-Rubeai, M. Selection methods for high-producing mammalian cell lines. *Trends Biotechnol.* **2007**, *25*, 425–432. [CrossRef] [PubMed]
24. Thomas, G. A new monoclonal antibody for the sensitive detection of atrazine with immunoassay in microtiter plate and dipstick format. *J. Agric. Food Chem.* **1993**, *41*, 1006–1011.
25. Shim, W.B.; Yang, Z.Y.; Kim, J.Y.; Choi, J.G.; Je, J.H.; Kang, S.J.; Kolosova, A.Y.; Eremin, S.A.; Chung, D.H. Immunochromatography using colloidal gold−antibody probe for the detection of atrazine in water samples. *J. Agric. Food Chem.* **2006**, *54*, 9728–9734. [CrossRef]
26. Sai, N.; Sun, W.; Wu, Y.; Sun, Z.; Yu, G.; Huang, G. A highly sensitive immunoassay for atrazine based on covalently linking the small molecule hapten to a urea–glutaraldehyde network on a polystyrene surface. *Int. Immunopharmacol.* **2016**, *40*, 480–486. [CrossRef]
27. Schlaeppi, J.M.; Foery, W.; Ramsteiner, K. Hydroxyatrazine and atrazine determination in soil and water by enzyme-linked immunosorbent assay using specific monoclonal antibodies. *J. Agric. Food Chem.* **1989**, *37*, 1532–1538. [CrossRef]

foods

MDPI

Article

Determination of Acetamiprid Residues in Vegetables by Indirect Competitive Chemiluminescence Enzyme Immunoassay

Zixin Zhu, Qiuyun Shi, Jianwei Wu, Kangli He, Jianguo Feng and Sa Dong *

School of Horticulture and Plant Protection, Yangzhou University, Yangzhou 225009, China
* Correspondence: dongsa.123@yzu.edu.cn

Abstract: Acetamiprid (ACE) is widely used in various vegetables to control pests, resulting in residues and posing a threat to human health. For the rapid detection of ACE residues in vegetables, an indirect competitive chemiluminescence enzyme immunoassay (ic-CLEIA) was established. The optimized experimental parameters were as follows: the concentrations of coating antigen (ACE-BSA) and anti-ACE monoclonal antibody were 0.4 and 0.6 μg/mL, respectively; the pre-incubation time of anti-ACE monoclonal antibody and ACE (sample) solution was 30 min; the dilution ratio of goat anti-mouse-HRP antibody was 1:2500; and the reaction time of chemiluminescence was 20 min. The half-maximum inhibition concentration (IC_{50}), the detection range (IC_{10}–IC_{90}), and the detection limit (LOD, IC_{10}) of the ic-CLEIA were 10.24, 0.70–96.31, and 0.70 ng/mL, respectively. The cross-reactivity rates of four neonicotinoid structural analogues (nitenpyram, thiacloprid, thiamethoxam, and clothianidin) were all less than 10%, showing good specificity. The average recovery rates in Chinese cabbage and cucumber were 82.7–112.2%, with the coefficient of variation (CV) lower than 9.19%, which was highly correlated with the results of high-performance liquid chromatography (HPLC). The established ic-CLEIA has the advantages of simple pretreatment and detection process, good sensitivity and accuracy, and can meet the needs of rapid screening of ACE residues in vegetables.

Keywords: acetamiprid; chemiluminescence enzyme immunoassay; rapid detection; residue; vegetable

Citation: Zhu, Z.; Shi, Q.; Wu, J.; He, K.; Feng, J.; Dong, S. Determination of Acetamiprid Residues in Vegetables by Indirect Competitive Chemiluminescence Enzyme Immunoassay. *Foods* **2022**, *11*, 2507. https://doi.org/10.3390/foods11162507

Academic Editor: Maojun Jin

Received: 3 August 2022
Accepted: 17 August 2022
Published: 19 August 2022

Publisher's Note: MDPI stays neutral with regard to jurisdictional claims in published maps and institutional affiliations.

Copyright: © 2022 by the authors. Licensee MDPI, Basel, Switzerland. This article is an open access article distributed under the terms and conditions of the Creative Commons Attribution (CC BY) license (https://creativecommons.org/licenses/by/4.0/).

1. Introduction

Acetamiprid (ACE), a new kind of chlorinated nicotinoid insecticide, with strong contact and stomach toxicity, as well as excellent internal absorption activity, is widely used in the control of aphid, whitefly, thrips and other pests on vegetables because of its quick insecticidal effect, low dosage, high activity, wide insecticidal spectrum and long duration [1,2]. Although ACE is a low-toxicity pesticide, current studies have shown that it has certain genotoxicity and cytotoxicity, has adverse effects on the nervous system and reproductive system of animals, and poses a threat to human health [3–9]. To limit its use, national and international organizations established maximum residue levels (MRLs) for ACE. In accordance with Annex II of Regulation (EC) No 396/2005, the MRLs of ACE in Chinese cabbage (code number: 243010) and cucumber (code number: 232010) were 1.5 and 0.3 mg/kg, respectively [10]. According to China's national standards (GB 2763-2021), the MRLs of ACE in different vegetables are 0.02–5 mg/kg (https://www.sdtdata.com/fx/fmoa/tsLibCard/183688.html, accessed on 18 May 2022). Therefore, it is of great significance to monitor ACE residual levels in vegetables to ensure food safety.

At present, the reported methods for detecting ACE residues are mainly focused on instrumental methods, such as high-performance liquid chromatography (HPLC), gas chromatography (GC), and chromatography–mass spectrometry (GC–MS) [11–15]. The instrument methods have high sensitivity, high accuracy, and good selectivity, but the equipment is expensive and requires professional laboratory personnel to operate, and the operation is complex and time-consuming, so it is not suitable for on-site rapid detection of ACE residues. The immunoassay method is simple, rapid, and can detect a large number of samples in a short time, which can be used for high-throughput detection of samples in the field [16]. At present, there are few studies on the use of chemiluminescence enzyme immunoassay for ACE detection, combining a highly specific immune response with a highly sensitive chemiluminescence reaction, which can increase the sensitivity by 2–3 fold compared with the traditional ELISA method [17].

In this study, a highly sensitive indirect competitive chemiluminescence enzyme immunoassay (ic-CLEIA) for ACE detection was established based on anti-ACE monoclonal antibody by optimizing the concentration of coating antigen and antibody, the pre-incubation time of anti-ACE monoclonal antibody and ACE (sample) solution, the dilution ratio of goat anti-mouse-HRP antibody and chemiluminescence reaction time. The ic-CLEIA was then evaluated by recovery experiment with simple sample pretreatment, which showed that the method was suitable for the detection of ACE in real samples. This study provides technical support for the rapid detection of ACE residues in vegetables, and has certain reference value for the detection of other pesticide residues.

2. Materials and Methods

2.1. Materials, Reagents and Equipment

Chinese cabbages and cucumbers (commercially available). Anti-ACE monoclonal antibody (anti-ACE mAb, 1.23 mg/mL) and coating antigen (ACE-BSA, 2.2 mg/mL) were acquired from Shandong Lvdu Bio-sciences and Technology Co., Ltd. (Binzhou, China). ACE standard (>99%) was purchased from Tiperi Instrument Equipment Co., Ltd. (Nanjing, China). Goat anti-mouse-HRP antibody was purchased from KPL Inc. (Gaithersburg, MD, USA). Skim milk were provided by Solarbio Science and Technology Co., Ltd., (Beijing, China). SuperSignal^{TM} ELISA Pico Chemiluminescent Substrate kit were purchased from Thermo Fisher Scientific (Thermo, Waltham, MA, USA). HPLC-grade acetonitrile was purchased from Tedia Company Inc. (Fairfield, OH, USA). The 96-well white plates were purchased from Corning Inc. (Corning, NY, USA). All other reagents and chemicals used were of analytical grade.

ACE standard solution: 0.05 g ACE standard was dissolved in 50 mL methanol to make a solution of 1 mg/mL. CBS solution: 2.93 g $NaHCO_3$ and 1.59 g Na_2CO_3 were weighed and dissolved in double distilled water, constant volume to 1 L, pH 9.6. PBS: 0.2 g KCl, 8.0 g NaCl, 2.9 g $Na_2HPO_4 \cdot 12H_2O$ and 0.2 g K_2HPO_4 were weighed and dissolved in double distilled water, constant volume to 1 L, pH 7.4. PBST: 0.5 mL Tween-20 was added to 1 L PBS. 3% MPBS: 0.3 g skim milk was weighed and dissolved in 10 mL PBS.

Electrothermal constant temperature precision incubator was purchased from Taisite Instrument Co., Ltd. (Tianjin, China). Multimode reader was purchased from Thermo Fisher Scientific (Thermo, Waltham, MA, USA). Highspeed freezing centrifuge was purchased from Eppendorf Inc. (Hamburg, Germany). LC-2000 high-performance liquid chromatograph was purchased from Hitachi Ltd. (Tokyo, Japan). Mili-Q ultrapure water machine was purchased from Millipore Ltd. (Burlington, MA, USA).

2.2. The Procedure of ic-CLEIA

First, a 96-well white plate was coated with the coating antigen (100 μL/well) in CBS at 37 °C for 2 h. After washing the plate with PBST (300 μL/well) three times, it was closed with 3% MPBS (200 μL/well) at 37 °C for 2 h. After washing the plate with PBST (300 μL/well) three times, the pre-incubation solution (diluted anti-ACE monoclonal antibody with PBS, mixed 50 μL antibody solution with 50 μL ACE solution, pre-incubated at 37 °C for a certain time) was added to the 96-well white plate and incubated at 37 °C for 1 h. After washing the plate with PBST (300 μL/well) three times, it was incubated with goat anti-mouse-HRP antibody (100 μL/well) in PBS at 37 °C for 1 h. Finally, equal proportions of chemiluminescence solution A and solution B in the SuperSignal^{TM} ELISA

Pico Chemiluminescent Substrate kit (100 μL/well) were added to the 96-well white plate. After incubation in the dark for a certain time, the luminescence value RLU was measured.

2.3. Optimization of ic-CLEIA

According to the previous results of indirect competitive enzyme-linked immunosorbent assay (ELISA) and checkerboard titration, the initial concentration of coating antigen was 0.37 μg/mL. The initial concentration of anti-ACE monoclonal antibody was 0.15 μg/mL. The pre-incubation time of anti-ACE monoclonal antibody and ACE (sample) solution was 20 min. The dilution ratio of goat anti-mouse HRP antibody was 1:5000. The chemiluminescence reaction time was 10 min.

Then, to improve the detection performance of ic-CLEIA, the effects of the coating antigen concentration (0.1, 0.2, 0.4, 0.8, and 1.6 μg/mL), antibody concentration (0.0375, 0.075, 0.15, 0.3, 0.6, and 1.2 μg/mL), the pre-incubation time of anti-ACE monoclonal antibody and ACE (sample) solution (10, 20, 30, and 40 min), the dilution ratio of goat anti-mouse-HRP antibody (1: 625, 1:1250, 1:2500, and 1:5000), and chemiluminescence reaction time (5, 10, 15, 20, 25, 30, and 35 min) on the sensitivity of ic-CLEIA were investigated by single factor experiment. According to the procedure of ic-CLEIA, standard curves for ACE detection using ic-CLEIA under each condition were established, and RLUmax (luminescence value without ACE) and the IC_{50} (half maximal inhibitory concentration, ACE concentration at 50% competitive inhibition) were calculated based on the standard curves. The RLUmax/IC_{50} ratio was used to evaluate the influence of specific factors on the detection performance of ic-CLIEA, and the higher the ratio, the higher the sensitivity under this condition [18]. The optimal reaction conditions of ic-CLEIA were high RLUmax/IC_{50}, a low IC_{50} and moderate RLUmax.

2.4. Establishment of the Standard Curve for ic-CLIEA

Under optimal conditions, the standard curve was drawn with ACE concentration (200, 100, 50, 25, 12.5, 6.25, 3.125, 1.5625, 0.8, 0.4, 0.2, and 0.1 ng/mL) as the abscissa and B/B0 (B is the RLU value with ACE and B0 is the RLU value without ACE) as the ordinate. Finally, the IC_{50}, the detection range (determined according to the standard curve), and the limit of detection (LOD, IC_{10}) of the ic-CLIEA method were calculated according to the standard curve.

2.5. Specificity of ic-CLIEA

The cross-reactivity rate was used to evaluate the specificity of ic-CLIEA method, and the higher the cross-reactivity rate, the worse the specificity. In this study, four neonicotinoid structural analogues (nitenpyram, thiacloprid, thiamethoxam, and clothianidin) were selected and determined by the established ic-CLIEA, then the IC_{50} and the cross-reactivity rates (CR%, CR% = IC_{50} of ACE/IC_{50} of analogue × 100) were calculated to evaluate the specificity of ic-CLIEA.

2.6. Sample Pretreatment

The Chinese cabbage and cucumber samples purchased from a supermarket were first confirmed by HPLC, and the samples without ACE residue were used for the recovery experiments.

The sample pretreatment used for ic-CLIEA detection was simplified based on previously reported methods [19]. Briefly, 100 g samples were homogenized and then the juice was squeezed out. A volume of 2 mL 99.5% acetone was added to the juice and left to stand for 5 min, then filtered with filter paper and the filtrate was collected. The filtrate was centrifuged at 6000 rpm for 5 min to obtain the supernatant, and the volume was fixed to 30 mL with sub-boiling water. The solution was passed through a 0.22 μm filter membrane and then determined by ic-CLIEA.

The sample pretreatment used for HPLC detection was performed according to the previously reported method with minor modification [20]. Briefly, 2 g samples were weighed and cut into 50 mL centrifuge tubes, then 10 mL acetonitrile was added to each centrifuge tube, and the samples were sonicated for 10 min. An appropriate amount of sodium chloride and anhydrous magnesium sulfate was added to the above mixture, and the mixture was vortexed and oscillated for 5 min, then centrifuged at 5000 rpm for 5 min. After centrifugation, 150 mg anhydrous magnesium sulfate and 50 mg primary secondary amine (PSA) were added to 1.5 mL of the above supernatant, thoroughly mixed by shaking, and centrifuged at 10,000 rpm for 5 min. The supernatant was passed through a 0.22 μm filter membrane and used for HPLC detection.

2.7. Elimination of Matrix Interference

A dilution method is usually used to eliminate matrix interference. A blank matrix without ACE is treated according to the sample pretreatment method during ic-CLIEA detection, and the filtrate is diluted with PBS and used to prepare ACE solution of serial concentration. Subsequently, ic-CLIEA was used for determination, and the standard curves were drawn, respectively. The influence of sample matrix on ACE detection was analyzed by comparing the above standard curve with the standard curve drawn by using PBS to prepare ACE solution without matrix.

2.8. Recovery Experiments

Since ACE is mostly used in Chinese cabbage, cabbage, cucumber and tomato, Chinese cabbage and cucumber were selected as actual samples in this experiment.

ACE standard solution was added to the blank samples to make the ACE content 1.5, 6, 30 μg/kg. After sample pretreatment, the established ic-CLIEA and HPLC methods were used for determination, and the recovery rate and the coefficient of variation (CV) were calculated.

HPLC conditions: Hypersil ODS (4.6 mm × 250 mm, 5 μm). The mobile phase was acetonitrile:water = 30:70 (V1:V2) and the flow rate of the mobile phase was 1.0 mL/min. The UV detection wavelength was 250 nm, the column temperature was 30 °C, and the injection volume was 5 μL.

2.9. Data Analysis

When the ic-CLIEA method was used for determination, the ACE concentration in the sample was calculated from the standard curve according to the RLU value, and then multiplied by the corresponding dilution ratio, which was the actual concentration of ACE in the sample. All experiments were repeated three times, and all the data in the results were the average values of the measured data.

3. Results

3.1. Optimization of Coating Antigen Concentration

Under the fixed concentration of antibody, the luminescence intensity and sensitivity increased with the increase in the coating antigen concentration, but when the concentration was too high, the steric hindrance increased, and the luminescence intensity and sensitivity decreased [18]. Therefore, it is necessary to optimize the concentration of coating antigen. The coating antigen was diluted to 0.1, 0.2, 0.4, 0.8, and 1.6 μg/mL and then determined by ic-CLIEA, respectively. The influence of the coating antigen concentration on the detection sensitivity and RLUmax/IC_{50} was analyzed. The results showed that RLUmax/IC_{50} increased firstly and then decreased with the increase in the coating antigen concentration. When the coating antigen concentration was 0.4 and 0.8 μg/mL, the values of RLUmax/IC_{50} were larger, and the IC_{50} gradually increased with the increase in coating antigen concentration. According to the principle that RLUmax/IC_{50} should be as large as possible and the IC_{50} should be as small as possible, the optimal concentration of coating antigen was selected as 0.4 μg/mL (Figure 1).

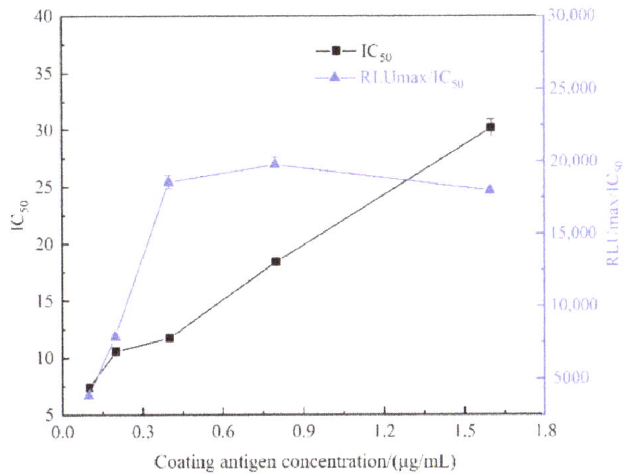

Figure 1. The IC_{50} and RLUmax/IC_{50} of ic-CLIEA under different coating antigen concentrations for ACE detection.

3.2. Optimization of Antibody Concentration

The 96-well white plate was coated with the optimal concentration of coating antigen, and anti-ACE monoclonal antibody was diluted to 0.0375, 0.075, 0.15, 0.3, 0.6, and 1.2 μg/mL, followed by ic-CLIEA assay. The influence of antibody concentration on the detection sensitivity and RLUmax/IC_{50} was analyzed. As shown in Figure 2, with the increase in antibody concentration, the RLUmax/IC_{50} increased first and then decreased. When the antibody concentration was 0.6 μg/mL, the RLUmax/IC_{50} reached the maximum, while the IC_{50} was small. Therefore, the optimal concentration of antibody was determined as 0.6 μg/mL.

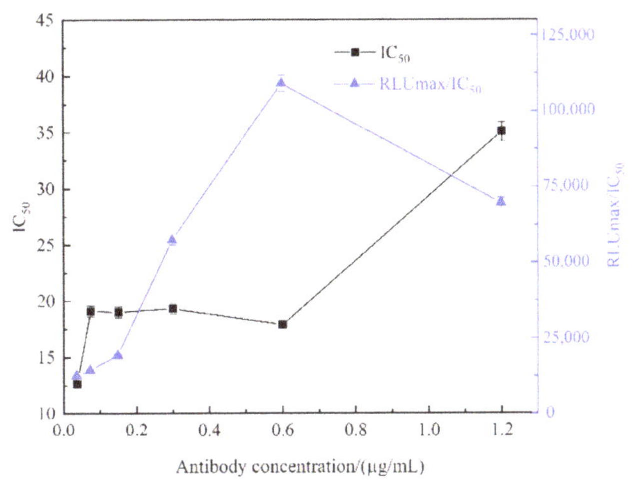

Figure 2. The IC_{50} and RLUmax/IC_{50} of ic-CLIEA with different antibody concentrations for ACE detection.

3.3. Optimization of Pre-Incubation Time

In ic-CLIEA detection, the full binding between antigen and antibody is related to the pre-incubation time of them, which has great impact on the sensitivity of the detection. Therefore, the pre-incubation time of anti-ACE monoclonal antibody and ACE solution should be optimized. As shown in Figure 3, when the pre-incubation time was 30 min,

RLUmax/IC$_{50}$ was largest, while the IC$_{50}$ was the smallest, so the optimal pre-incubation time between ACE and its mAb was determined to be 30 min.

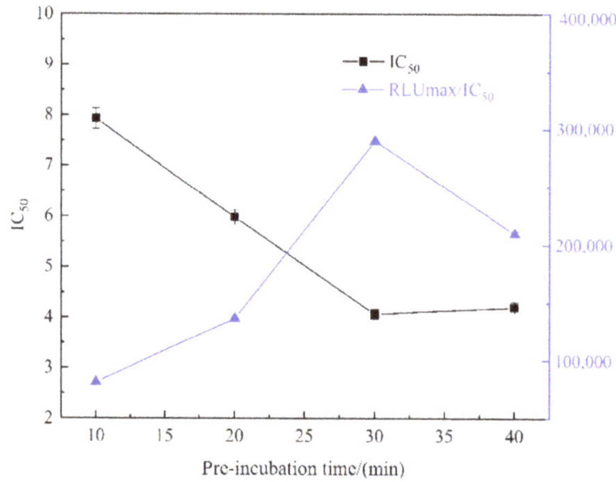

Figure 3. The IC$_{50}$ and RLUmax/IC$_{50}$ of ic-CLIEA with different pre-incubation time between ACE and its mAb for ACE detection.

3.4. Optimization of the Dilution Ratio of Goat Anti-Mouse-HRP Antibody

The effect of goat anti-mouse HRP secondary antibody concentration on the sensitivity of ic-CLEIA was further investigated. Goat anti-mouse HRP secondary antibody (1 mg/mL) was diluted by 1:625, 1:1250, 1:2500, and 1:5000, respectively, and then determined by ic-CLIEA. As shown in Figure 4, with the increase in the dilution ratio of goat anti-mouse HRP secondary antibody, the RLUmax/IC$_{50}$ increased first and then decreased, while the IC$_{50}$ gradually decreased. When the dilution ratio of the secondary antibody was 1:2500, the RLUmax/IC$_{50}$ was the maximum, while the IC$_{50}$ was small. Therefore, dilution of 1:2500 was selected as the optimal dilution ratio of goat anti-mouse HRP secondary antibody.

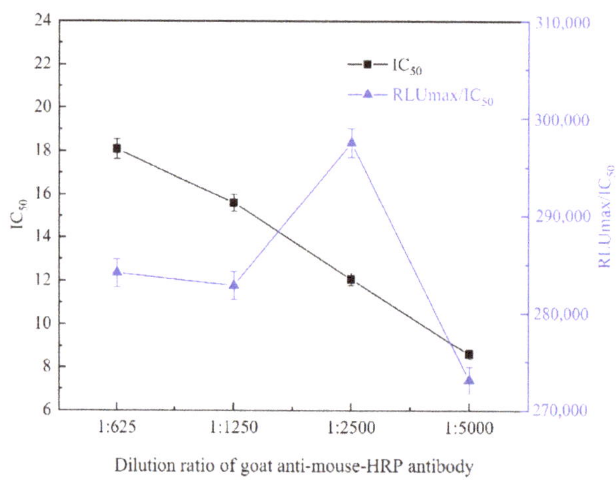

Figure 4. The IC$_{50}$ and RLUmax/IC$_{50}$ of ic-CLIEA with different goat anti-mouse-HRP antibody dilution ratios for ACE detection.

3.5. Optimization of Chemiluminescence Reaction Time

In ic-CLIEA detection, the luminescence intensity first increased and then decreased with the extension of time after the addition of luminescent substrate, thus the optimal reaction time of chemiluminescence must be selected to ensure the sensitivity and accuracy of the experiment [18]. The reaction time of chemiluminescence substrate was set as 10, 15, 20, 25, 30, and 35 min, respectively, and the luminescence intensity of each group was measured. The results showed that the luminescence intensity reached the maximum value at 20 min, so the optimal reaction time of chemiluminescence was determined to be 20 min (Figure 5).

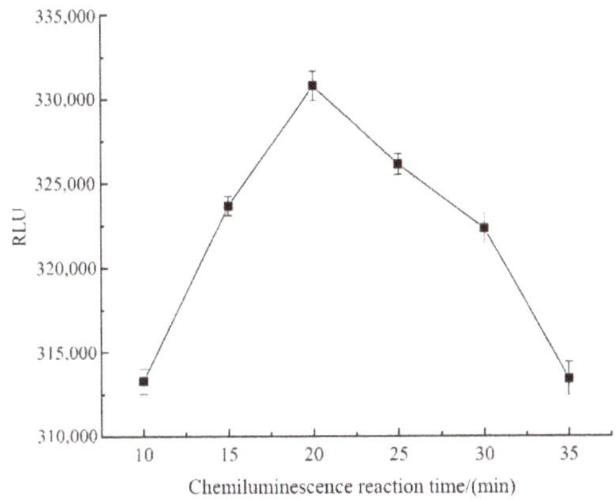

Figure 5. RLU values at different chemiluminescence reaction time.

3.6. The Standard Curve of ic-CLIEA

Based on the optimal ic-CLIEA detection conditions, the ACE concentration was taken as the abscissa and the B/B0 value as the ordinate, and the standard curve of ACE was drawn and fitted by Origin 2018. As shown in Figure 6, the standard curve equation was $y = 7.11 + 107.09/[1 + (x/12.09)^{0.80}]$, $R^2 = 0.995$. According to the standard curve, the IC_{50} was 10.24 ng/mL, the detection range (determined as the IC_{10}–IC_{90} according to the standard curve) was 0.70–96.31 ng/mL, and the LOD (IC_{10}) was 0.70 ng/mL. According to the pretreatment and dilution method in this study, the LOD of ACE in actual sample was calculated to be 1.26 μg/kg.

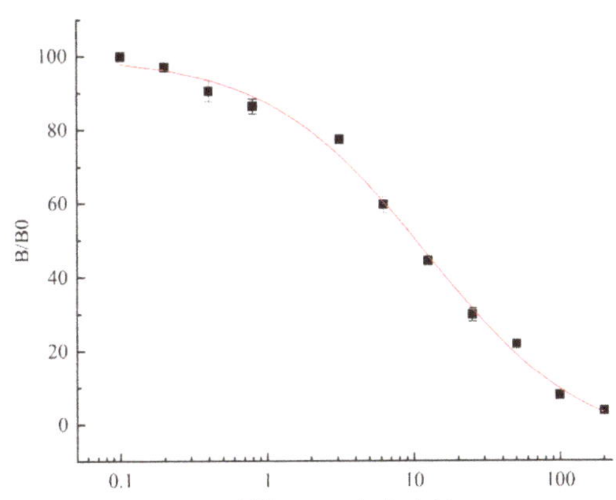

Figure 6. Standard curve of ic-CLIEA for ACE detection.

3.7. Specificity of ic-CLIEA

Four structural analogues of neonicotinoids were selected for cross-reactivity determination, and the results were shown in Table 1. The cross-reactivity rates of ic-CLIEA for nitenpyram, thiacloprid, thiamethoxam, and clothianidin were all less than 10%, indicating that the ic-CLIEA had good specificity.

Table 1. Cross-reactivity of ACE and its analogues with anti-ACE mAb determined by ic-CLIEA (n = 3).

Pesticide	Structural Formula	IC_{50} (ng/mL)	Cross-Reactivity
acetamiprid	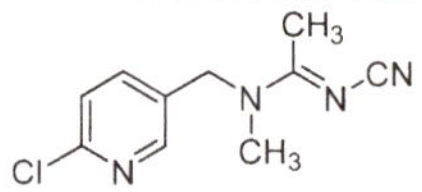	10.24	100
nitenpyram	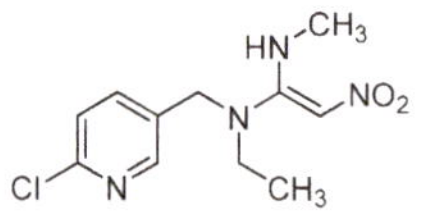	522.45	1.96
thiacloprid	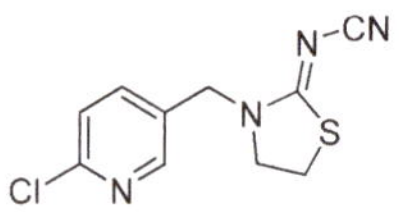	213.78	4.79
thiamethoxam	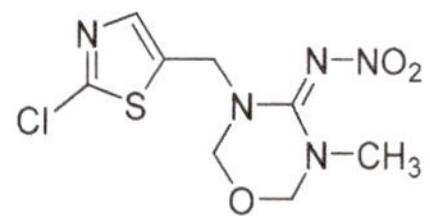	>10,000	<0.1
clothianidin	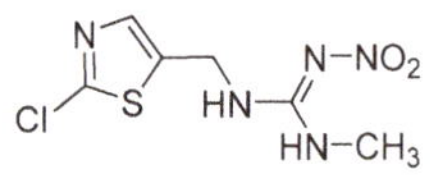	118.66	8.63

3.8. Elimination of Matrix Interference

In the process of immunoassay, it is crucial to eliminate matrix interference, because the pH, ionic strength, and organic matter content of real samples will interfere with the specific reaction between antigen and antibody, thus affecting the sensitivity of detection [21]. In this experiment, the extract juice of Chinese cabbage and cucumber were diluted 0, 4, 6, and 8 fold with PBS solution, respectively. Subsequently, standard curves drawn with the diluted solution of the matrix juice were compared with those drawn with PBS, and the appropriate dilution ratio was chosen to eliminate matrix interference. Figure 7 shows that the extract juice of Chinese cabbage and cucumber has little influence on the standard curve after 6-fold dilution, so the 6-fold dilution of the sample extract juice was chosen for the determination of ACE in the subsequent test.

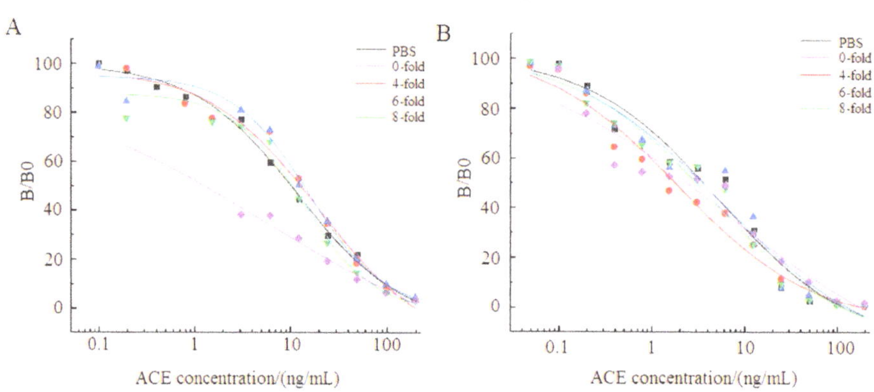

Figure 7. Matrix effect of Chinese cabbage (**A**) and cucumber (**B**) after dilution 0, 4, 6, and 8 fold with PBS solution.

3.9. Recovery Experiments

According to the LOD (the LOD in real samples was 1.26 μg/kg) and the detection range (the detection range in real samples was 1.26–173.36 μg/kg) of the ic-CLIEA, the added concentration of ACE in real samples was set as 1.5, 6, and 30 μg/kg. As shown in Table 2, the average recovery rate of ACE in Chinese cabbage and cucumber determined by ic-CLIEA was 82.7–112.2% with CV less than 9.19%, and the average recovery rate of ACE determined by HPLC was 80.7–118.00% with CV less than 9.08%. The results showed that the established ic-CLIEA method was accurate and reliable, and had a good correlation with HPLC, which could be used for the detection of ACE residue in vegetables.

Table 2. Recoveries of acetamiprid in Chinese cabbage and cucumber (n = 3).

Sample	Spiked Level/(μg/kg)	ic-CLIEA			HPLC		
		Found/(μg/kg)	Average Recovery/%	CV/%	Found/(μg/kg)	Average Recovery/%	CV%
Chinese cabbage	1.5	1.24	82.7	2.71	1.77	118.0	4.97
	6	5.54	92.3	5.68	6.70	111.7	9.08
	30	33.67	112.2	8.48	31.71	105.7	6.10
Cucumber	1.5	1.29	86.0	9.19	1.21	80.7	3.21
	6	5.14	85.7	4.01	4.86	81.0	2.01
	30	29.78	99.3	4.20	28.89	96.3	0.80

3.10. Comparison of Some Published Results for ACE Rapid Detection

In recent years, many rapid methods for ACE residue detection have been reported. Comparing the results of this research with the published results, it was shown that the ic-CLIEA method established in this study does not require the synthesis of any materials, the detection materials and reagents are easily available, the pretreatment and operation procedures are simple, as well as has a wide linear range and a low detection limit, which is suitable for ACE detection (Table 3).

Table 3. Comparison of some published results for ACE detection with this research.

Methods	Synthesis of Materials	Linear Range (ng/mL)	LOD (ng/mL)
lateral flow assay [22]	no need	no	1
lateral flow assay [23]	AuNPs@polyA-cDNA	no	0.33
colorimetric [24]	gold nanoparticles (AuNPs)	5567–66803	848
surface-enhanced Raman [25]	Ag-coated cellophane	no	1000
colorimetric and fluorescence [26]	AuNPs	5.56–222	0.08
fluorescence [27]	Cationic carbon dots (cCDs)	0.357–26.8	0.067
chemiluminescence sensor [28]	graphene oxide (GO) and AuNPs	0.0047–2	0.002
this work	no need	0.70–96.31	0.70

4. Discussion

In the immunoassay, the sensitivity and stability of the method mainly depend on the balance of the specific reaction and reversibility reaction between antigen and antibody [29]. Therefore, a series of conditions affecting the sensitivity of ic-CLIEA were optimized in this study, including the concentration of coating antigen, antibody concentration, the pre-incubation time of anti-ACE monoclonal antibody and ACE (sample) solution, the dilution ratio of goat anti-mouse HRP antibody, and chemiluminescence reaction time. The results showed that the concentrations of antigen and antibody were the key factors affecting the reaction balance. If the concentrations of antigen and antibody are too low, the reaction is not complete, and if the concentrations of antigen and antibody are too high, it is easy to cause multilayer adsorption, which leads to the mutual cover of the antigenic determinants, thus affecting the stability and sensitivity of the analytical method. In addition, the reaction time of antigen and antibody is also an important factor affecting the sensitivity and accuracy of the method. With the extension of the reaction time, the IC_{50} of the method shows a trend of decreasing first and then increasing, which may be because too short a reaction time will lead to incomplete binding between antigen and antibody, and too long a reaction time will easily cause non-specific adsorption. Therefore, only appropriate reaction time can make the sensitivity of the method reach the best. Matrix effect refers to the non-specific reaction to the substance in the extract during immunoassay, which may lead to inaccurate results in actual sample analysis. Usually, matrix effects can be eliminated by simple dilution prior to analysis. The dilution ratio at which there is no significant difference between the absorbance of the extract solution with or without sample matrix should be confirmed to manage the matrix effect [30].

Finally, a highly sensitive ic-CLIEA for ACE detection was successfully established, with an IC_{50} of 10.24 ng/mL, a detection range (IC_{10}–IC_{90}) of 0.70–96.31 ng/mL, and a LOD (IC_{10}) of 0.70 ng/mL (according to the pretreatment and dilution method in this study, the LOD in real samples was 1.26 μg/kg). Although the LOD of the established ic-CLIEA in this study is not the lowest among all reported methods, the materials and reagents used in this method are easily available, the accuracy and selectivity are high, the detection sensitivity can meet the MRL requirements of ACE in vegetables, and the quantitative detection can be achieved. Most importantly, the pretreatment method is very simple and suitable for rapid high-throughput screening of ACE residues in vegetables, which has a good application prospect. In addition, the detection method established in this study is also suitable for the rapid detection of other compounds and pesticides, but the experimental parameters need to be re-optimized to improve the detection sensitivity.

Author Contributions: Conceptualization, J.F. and S.D.; data curation, Z.Z., Q.S. and J.W.; formal analysis, Z.Z. and K.H.; funding acquisition, S.D.; investigation, S.D.; methodology, Z.Z. and Q.S.; project administration, S.D.; resources, S.D.; software, Q.S.; supervision, S.D.; validation, Z.Z., Q.S. and J.W.; visualization, K.H.; writing—original draft, Z.Z. and Q.S.; writing—review and editing, Z.Z., Q.S. and S.D. All authors have read and agreed to the published version of the manuscript.

Funding: This research was funded by Natural Science Foundation of Jiangsu Province, grant number BK20180916.

Institutional Review Board Statement: Not applicable.

Informed Consent Statement: Not applicable.

Data Availability Statement: The data presented in this study are available on request from the corresponding author.

Conflicts of Interest: The authors declare no conflict of interest.

References

1. Wang, A.; Mahai, G.; Wan, Y.; Jiang, Y.; Meng, Q.; Xia, W.; He, Z.; Xu, S. Neonicotinoids and carbendazim in indoor dust from three cities in China: Spatial and temporal variations. *Sci. Total Environ.* **2019**, *695*, 133790. [CrossRef] [PubMed]
2. Taillebois, E.; Alamiddine, Z.; Brazier, C.; Graton, J.; Laurent, A.D.; Thany, S.H.; Le Questel, J.Y. Molecular features and toxicological properties of four common pesticides, acetamiprid, deltamethrin, chlorpyriphos and fipronil. *Bioorg. Med. Chem.* **2015**, *23*, 1540–1550. [CrossRef] [PubMed]
3. Bagri, P.; Jain, S.K. Assessment of acetamiprid-induced genotoxic effects in bone marrow cells of Swiss albino male mice. *Drug Chem. Toxicol.* **2019**, *42*, 357–363. [CrossRef] [PubMed]
4. Ma, X.; Li, H.; Xiong, J.; Mehler, W.T.; You, J. Developmental Toxicity of a Neonicotinoid Insecticide, Acetamiprid to Zebrafish Embryos. *J. Agric. Food Chem.* **2019**, *67*, 2429–2436. [CrossRef]
5. Terayama, H.; Qu, N.; Endo, H.; Ito, M.; Tsukamoto, H.; Umemoto, K.; Kawakami, S.; Fujino, Y.; Tatemichi, M.; Sakabe, K. Effect of acetamiprid on the immature murine testes. *Int. J. Environ. Health Res.* **2018**, *28*, 683–696. [CrossRef]
6. Çamlıca, Y.; Bediz, S.C.; Çömelekoğlu, Ü.; Yilmaz, Ş.N. Toxic effect of acetamiprid on Rana ridibunda sciatic nerve (electrophysiological and histopathological potential). *Drug Chem. Toxicol.* **2019**, *42*, 264–269. [CrossRef]
7. Mosbah, R.; Djerrou, Z.; Mantovani, A. Protective effect of Nigella sativa oil against acetamiprid induced reproductive toxicity in male rats. *Drug Chem. Toxicol.* **2018**, *41*, 206–212. [CrossRef]

8. Baysal, M.; Atlı-Eklioğlu, Ö. Comparison of the toxicity of pure compounds and commercial formulations of imidacloprid and acetamiprid on HT-29 cells: Single and mixture exposure. *Food Chem. Toxicol. Int. J. Publ. Br. Ind. Biol. Res. Assoc.* **2021**, *155*, 112430. [CrossRef]
9. Saggioro, E.M.; do Espírito Santo, D.G.; Sales Júnior, S.F.; Hauser-Davis, R.A.; Correia, F.V. Lethal and sublethal effects of acetamiprid on Eisenia andrei: Behavior, reproduction, cytotoxicity and oxidative stress. *Ecotoxicol. Environ. Saf.* **2019**, *183*, 109572. [CrossRef]
10. EFSA (European Food Safety Authority); Bellisai, G.; Bernasconi, G.; Brancato, A.; Carrasco Cabrera, L.; Ferreira, L.; Giner, G.; Greco, L.; Jarrah, S.; Kazocina, A.; et al. Modification of the existing maximum residue levels for acetamiprid in various crops. *EFSA J.* **2021**, *19*, e06830. [CrossRef]
11. Lee, J.; Kim, B.J.; Kim, E.; Kim, J.H. Dissipation kinetics and the pre-harvest residue limits of acetamiprid and chlorantraniliprole in kimchi cabbage using ultra-performance liquid chromatography-tandem mass spectrometry. *Molecules* **2019**, *24*, 2616. [CrossRef]
12. Mohapatra, S.; Siddamallaiah, L.; Matadha, N.Y. Behavior of acetamiprid, azoxystrobin, pyraclostrobin, and lambda-cyhalothrin in/on pomegranate tissues. *Environ. Sci. Pollut. Res. Int.* **2021**, *28*, 27481–27492. [CrossRef]
13. Yıldırım, İ.; Çiftçi, U. Monitoring of pesticide residues in peppers from Çanakkale (Turkey) public market using QuEChERS method and LC-MS/MS and GC-MS/MS detection. *Environ. Monit. Assess.* **2022**, *194*, 570. [CrossRef]
14. Pietrzak, D.; Wątor, K.; Pękała, D.; Wójcik, J.; Chochorek, A.; Kmiecik, E.; Kania, J. LC-MS/MS method validation for determination of selected neonicotinoids in groundwater for the purpose of a column experiment. *J. Environ. Sci. Health. Part B Pestic. Food Contam. Agric. Wastes* **2019**, *54*, 424–431. [CrossRef]
15. Srivastava, A.; Chabra, A.; Singh, G.P.; Srivastava, P.C. Efficacy of different decontamination processes in mitigation of pesticide residues from chili crop. *J. Food Prot.* **2021**, *84*, 767–771. [CrossRef]
16. Xu, C.; Peng, C.; Hao, K.; Jin, Z.; Wang, W. Chemiluminescence enzyme immunoassay (CLEIA) for the determination of chloramphenicol residues in aquatic tissues. *Luminescence* **2006**, *21*, 126–128. [CrossRef]
17. Xiao, Z.; Tian, W.; Xu, Z.; Sun, Y. Development of a chemiluminescent enzyme immunoassay to detect cadmium in cereal samples. *Food Agric. Immunol.* **2018**, *29*, 1184–1196. [CrossRef]
18. Yang, W.; Dong, J.; Shen, Y.; Yang, J.; Wang, H.; Xu, Z.; Yang, X.; Sun, Y. Stablishment of chemiluminescence enzyme immunoassay for furatone metabolites in shrimp. *Chin. J. Anal. Chem.* **2012**, *40*, 1816–1821. [CrossRef]
19. Li, X.; Wu, F.; Pang, Y.; Nie, H.; Huang, S.; Luo, Y. Detection of pesticide residue acetamiprid by SERS based on aptamers regulating carbon dots catalytic reactions. *Sci. Technol. Food Ind.* **2021**, *42*, 236–244. [CrossRef]
20. Chen, L.; Wang, J.; Du, P.; Tang, X.; Zhao, K.; Pan, C. Microwave treatment of imidacloprid and acetamidine residues in leek-purification by inverse solid phase dispersion and detection by liquid chromatography. *Chin. J. Anal. Chem.* **2008**, *36*, 1364–1368. [CrossRef]
21. Xu, Z.L.; Sun, W.J.; Yang, J.Y.; Jiang, Y.M.; Campbell, K.; Shen, Y.D.; Lei, H.T.; Zeng, D.P.; Wang, H.; Sun, Y.M. Development of a solid-phase extraction coupling chemiluminescent enzyme immunoassay for determination of organophosphorus pesticides in environmental water samples. *J. Agric. Food Chem.* **2012**, *60*, 2069–2075. [CrossRef] [PubMed]
22. Liu, L.; Suryoprabowo, S.; Zheng, Q.; Song, S.; Kuang, H. Development of an immunochromatographic strip for detection of acetamiprid in cucumber and apple samples. *Food Agric. Immunol.* **2017**, *28*, 767–778. [CrossRef]
23. Mao, M.; Xie, Z.; Ma, P.; Peng, C.; Wang, Z.; Wei, X.; Liu, G. Design and optimizing gold nanoparticle-cDNA nanoprobes for aptamer-based lateral flow assay: Application to rapid detection of acetamiprid. *Biosens. Bioelectron.* **2022**, *207*, 114114. [CrossRef] [PubMed]
24. Xu, C.; Lin, M.; Song, C.; Chen, D.; Bian, C. A gold nanoparticle-based visual aptasensor for rapid detection of acetamiprid residues in agricultural products using a smartphone. *RSC Adv.* **2022**, *12*, 5540–5545. [CrossRef]
25. Wei, W.; Huang, Q. Preparation of cellophane-based substrate and its SERS performance on the detection of CV and acetamiprid. *Spectrochim. Acta. Part A Mol. Biomol. Spectrosc.* **2018**, *193*, 8–13. [CrossRef]
26. Yang, L.; Sun, H.; Wang, X.; Yao, W.; Zhang, W.; Jiang, L. An aptamer based aggregation assay for the neonicotinoid insecticide acetamiprid using fluorescent upconversion nanoparticles and DNA functionalized gold nanoparticles. *Mikrochim. Acta* **2019**, *186*, 308. [CrossRef]
27. Saberi, Z.; Rezaei, B.; Ensafi, A.A. Fluorometric label-free aptasensor for detection of the pesticide acetamiprid by using cationic carbon dots prepared with cetrimonium bromide. *Mikrochim. Acta* **2019**, *186*, 273. [CrossRef]
28. Xiu, F.; Lu, Y.; Qi, Y.; Wang, Y.; He, J. Ultrasensitive and practical chemiluminescence sensing pesticide residue acetamiprid in agricultural products and environment: Combination of synergistically coupled co-amplifying signal and smart interface engineering. *Talanta* **2021**, *235*, 122811. [CrossRef]
29. Zhu, F.; Li, H.; Wang, H.; Yang, J.; Xu, Z.; Lei, H.; Xiao, Z.; Sun, Y.; Shen, Y. Development of indirect competitive chemiluninescene enzyme immunoassay (icELISA) method for deternination of iso-tenuazonic acid in fruit and tomato sauce. *Food Sci.* **2017**, *38*, 250–254. [CrossRef]
30. Dong, S.; Zhang, C.; Liu, Y.; Zhang, X.; Xie, Y.; Zhong, J.; Xu, C.; Liu, X. Simultaneous production of monoclonal antibodies against *Bacillus thuringiensis* (Bt) Cry1 toxins using a mixture immunization. *Anal. Biochem.* **2017**, *531*, 60–66. [CrossRef]

foods MDPI

Communication

Quantum-Dot-Bead-Based Fluorescence-Linked Immunosorbent Assay for Sensitive Detection of Cry2A Toxin in Cereals Using Nanobodies

Yulou Qiu, Ajuan You, Xianshu Fu, Mingzhou Zhang, Haifeng Cui, Biao Zhang, Weiwei Qin, Zihong Ye and Xiaoping Yu *

Zhejiang Provincial Key Laboratory of Biometrology and Inspection & Quarantine, College of Life Science, China Jiliang University, Hangzhou 310018, China
* Correspondence: yxp@cjlu.edu.cn

Abstract: In this study, a quantum-dot-bead (QB)-based fluorescence-linked immunosorbent assay (FLISA) using nanobodies was established for sensitive determination of the Cry2A toxin in cereal. QBs were used as the fluorescent probe and conjugated with a Cry2A polyclonal antibody. An anti-Cry2A nanobody P2 was expressed and used as the capture antibody. The results revealed that the low detection limit of the developed QB-FLISA was 0.41 ng/mL, which had a 19-times higher sensitivity than the traditional colorimetric ELISA. The proposed assay exhibited a high specificity for the Cry2A toxin, and it had no evident cross-reactions with other Cry toxins. The recoveries of Cry2A from the spiked cereal sample ranged from 86.6–117.3%, with a coefficient of variation lower than 9%. Moreover, sample analysis results of the QB-FLISA and commercial ELISA kit correlated well with each other. These results indicated that the developed QB-FLISA provides a potential approach for the sensitive determination of the Cry2A toxin in cereals.

Keywords: Cry2A; nanobodies; QBs; FLISA

Citation: Qiu, Y.; You, A.; Fu, X.; Zhang, M.; Cui, H.; Zhang, B.; Qin, W.; Ye, Z.; Yu, X. Quantum-Dot-Bead-Based Fluorescence-Linked Immunosorbent Assay for Sensitive Detection of Cry2A Toxin in Cereals Using Nanobodies. *Foods* **2022**, *11*, 2780. https://doi.org/10.3390/foods11182780

Academic Editors: Amin Mousavi Khaneghah and Maojun Jin

Received: 2 July 2022
Accepted: 25 August 2022
Published: 9 September 2022

Publisher's Note: MDPI stays neutral with regard to jurisdictional claims in published maps and institutional affiliations.

Copyright: © 2022 by the authors. Licensee MDPI, Basel, Switzerland. This article is an open access article distributed under the terms and conditions of the Creative Commons Attribution (CC BY) license (https://creativecommons.org/licenses/by/4.0/).

1. Introduction

Cry toxins are a group of parasporal crystal proteins produced by *Bacillus thuringiensis* during its sporulation phase, exhibiting excellent insecticidal activity [1–3]. At present, its mechanism of action is still unclear. The main proposed mechanism is that Cry toxins combine with the receptors on the midgut cell membrane of insects, forming dissolved pores and resulting in cell lysis, eventually leading to insect death [4,5]. Due to their high toxicity and specificity to insects, Cry-toxin-based insecticides are still the most commonly used biological insecticides since the commercial Cry-toxin-based insecticides were firstly produced in 1938 [6,7]. In addition, Cry toxins are widely introduced into genetically modified (GM) crops for protection against insect pests [8,9]. However, the safety of Cry toxins has not been universally recognized and accepted, and the widespread use of Cry toxins may pose potential threats to the eco-environment and public health [10,11]. Many countries have established mandatory labeling policies for GM products with a threshold level of 0–5% [12]. Therefore, detecting and quantifying Cry toxins in agricultural products and the environment are necessary and important.

Various methodologies, including bioassays, mass spectrometry, surface plasmon resonance (SPR) biosensors, electrochemical immunosensors, real-time PCR, and enzyme-linked immunosorbent assays (ELISAs), have been established for Cry toxins' analysis [12–15]. The ELISA methods are appropriate for the on-site determination of Cry toxins with the advantages of being convenient, rapid, and cost efficient [16]. Among them, the most frequently used format is double-antibody sandwich ELISAs (DAS-ELISAs), which generally depend on the traditional polyclonal antibodies (PAbs) and monoclonal antibodies (MAbs) [17,18]. Besides, genetically engineered antibodies (e.g., ScFvs) and phage-displayed peptides have been applied in DAS-ELISA for Cry toxin analysis [19,20]. However, ScFvs and peptides usually exhibited relatively low affinity and poor stability [21,22].

Nanobodies, which were first discovered by Hamers-Casterman in 1993, have attracted much attention in recent years [23–26]. They are a class of single-domain antibodies derived from heavy-chain antibodies of camelid or related species. Nanobodies are the smallest known functional antibodies, having a low molecular weight of approximately 15 kDa and a small size of 2.5 × 4 nm [27]. Due to their small size and unique structure, nanobodies exhibit many advantageous characteristics, including the ease of manipulation, the ability to recognize inaccessible epitopes, high affinity, and high stability.

Besides, the sensitivity of immunoassays is greatly affected by the signal probes. To date, many fluorescent probes (e.g., quantum dots (QDs), up-conversion nanoparticles, and lanthanide ions) instead of traditional enzymes have been applied to enhance the sensitivity of immunoassays [28–35]. Among them, QDs are ideal fluorescent materials because of their narrow emission spectra, broad excitation, high fluorescent intensity, and high stability [36,37]. In addition, QBs are prepared by embedding plenty of QDs in a polymer matrix, and they show stronger fluorescent intensity than QDs. Li et al. [38] established an immunochromatographic assay using QBs (QBs-ICA) for the detection of cyproheptadine hydrochloride with an IC_{50} of 1.38 ng/mL. The sensitivity of QBs-ICA was 10-times higher than previously report. Qie et al. [39] developed a CdSe/ZnS-QB-based lateral flow immunoassay (LFI) for the T-2 toxin with an LOD of 10 fg/mL, showing an eight-fold higher sensitivity than traditional LFI system.

In our previous work, several phage-displayed nanobodies against the Cry2A toxin were isolated from a naive nanobody library [40]. In this study, the anti-Cry2A nanobody P2 was expressed and purified as the capture antibody. QBs were used to couple with the anti-Cry2A PAb to serve as the fluorescent probe. Then, a sensitive QB-based sandwich fluorescence-linked immunosorbent assay (FLISA) using nanobodies was established for the analysis of the Cry2A toxin (Figure 1). The sensitivity and selectivity of the established assay were investigated. In addition, the QB-FLISA and commercial ELISA kit were used to detect the Cry2A toxin in cereal samples. The established QB-FLISA has the potential to be a reliable method for detecting the Cry2A toxin in cereals.

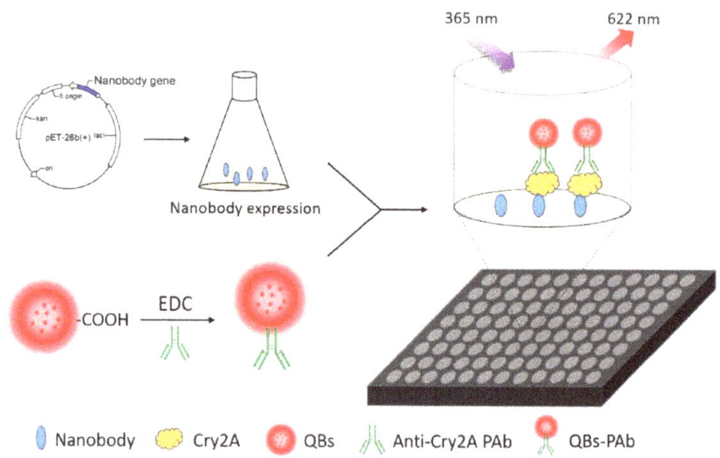

Figure 1. Schematic diagram of the nanobody-based QB-FLISA.

2. Materials and Methods

2.1. Chemicals and Reagents

Cry toxins (Cry2A, Cry1Ab, Cry1B, and Cry3Bb) and the Cry2A commercial ELISA kit were purchased from You Long Bio. Co., Ltd. (Shanghai, China). QBs were purchased from Kundao (Shanghai, China). Phage-displayed nanobody P2 and anti-Cry2A PAb were prepared previously [40]. HRP-labeled goat anti-rabbit antibody was purchased from Solarbio (Beijing, China). Restriction enzymes *Not* I, *Nco* I, T4 DNA ligase, and TaqDNA polymerase were purchased from Takara Co. (Dalian, China). Isopropylthio-β-D-galactoside (IPTG), 96-well microplates, and nickel-nitrilotriacetic acid (Ni-NTA) resin were purchased from Sangon Biotech (Shanghai, China). Bovine serum albumin (BSA) was purchased from Sigma-Aldrich (St. Louis, MO, USA). The TMB substrate was purchased from Beyotime (Shanghai, China).

2.2. Expression and Purification of Anti-Cry2A Nanobodies

A gene fragment of the anti-Cry2A nanobody P2 was amplified by PCR (94 °C for 3 min, followed by 30 cycles of 94 °C for 30 s, 55 °C for 30 s, and 72 °C for 30 s) and digested by restriction enzymes *Not* I and *Nco* I. Then, the gene fragment of the nanobody P2 was gel purified and inserted into the pET-26b(+) vector. The recombinant vectors were transferred into *E. coli* DH5α-competent cells. After confirmation was achieved by PCR and sequencing, the target recombinant vector was transferred into *E. coli* Rosetta (DE3). The recombinant *E. coli* strains were spread on the LB plate supplemented with 40 μg/mL kanamycin, and individual colonies were incubated in LB liquid medium by shaking overnight. Then, the cultured bacteria were inoculated into 100 mL of fresh LB-kanamycin liquid culture and incubated to an OD600 of 0.5. The recombinant cells were induced with IPTG at 0.2 mM by shaking (200 rpm) for 12 h and harvested by centrifugation (8000× *g*, 15 min). Subsequently, the cell suspensions were broken by sonication and separated by centrifugation (10,000× *g*, 15 min). Then, supernatants were collected and filtered through a 0.22 μm membrane. The soluble nanobody with a 6 × His-tag was purified by the Ni-NTA affinity column. The purified nanobodies were confirmed by SDS-PAGE in accordance with the standard protocol. The concentration of purified nanobodies was measured using a Nanodrop 1000 (Thermo, Waltham, MA, USA).

2.3. Preparation of QBs-PAb

QBs-PAb conjugates were synthesized according to the published procedure with some modification [41]. In brief, 10 μL of QBs (10 mg/mL) and 5 μL of EDC (10 mg/mL) were added to 500 μL of PB buffer (10 mM, pH 6.0) and mixed at RT for 30 min. After these reaction mixtures were separated by centrifugation (10,000× *g*, 10 min), the precipitate was collected and resuspended in 400 μL PB buffer. Subsequently, 50 μL of anti-Cry2A PAbs (2.0 mg/mL) was added into the buffer and mixed at RT for another 30 min. Afterwards, the mixture was blocked by 1% BSA solution for 1 h incubation. After centrifugation (10,000× *g*, 10 min), the precipitate was resuspended in 10 mM PBS buffer with 1% BSA and 0.5% Tween-20. Finally, the QBs-PAb conjugates were stored at 4 °C for further use. Transmission electron microscopy (TEM) images were recorded by an electron microscope (JEOL, Tokyo, Japan) operated at 200 kV. Fluorescence spectra (excitation at 365 nm) were recorded with a fluorescence microplate reader (Tecan, Männedorf, Switzerland). Dynamic light scattering (DLS) analysis was carried out on a particle size analyzer (Malvern Instruments, Malvern, UK).

2.4. QBs-Based FLISA

Using nanobodies as the capture antibody, a double-antibody sandwich FLISA (DAS-FLISA) based on QBs for detecting the Cry2A toxin was developed. First, the nanobody P2 (5.0 μg/mL, 100 μL/well) was added to the black microplate. After incubation overnight at 4 °C, the plates were washed three times in 0.05% PBST, and the excess binding sites were blocked for 2 h in blocking buffer (3% BSA in 0.1% PBST). Subsequently, the microplate was washed three times with 0.05% PBST, and different concentrations of Cry2A toxins (0–1000 ng/mL, 100 μL/well) were added to the washed wells for incubation at 37 °C for 1 h. Next, QBs-PAb probes were added to the microplate and incubated for 1 h. Finally, unbound QBs-PAb probes were discarded by washing six times, and the fluorescence intensity of each well was measured by a multimode microplate reader (Tecan, Männedorf, Switzerland). The working concentrations of the nanobody P2 and QBs-PAb were optimized in advance to obtain the best performance of the QB-FLISA.

2.5. Cross-Reactivity

The specificity of the QB-FLISA was evaluated against the Cry1Ab, Cry1B, and Cry3Bb toxins. Briefly, 100 μL of the nanobody P2 per well was added to black microtiter plates and incubated overnight at 4 °C. After the plates were blocked with blocking buffer and washed thrice with 0.05% PBST, 100 μL of different Cry toxins (1000 ng/mL) was added into the microtiter plates for 1 h incubation. The following steps were the same as those described above. Each assay was performed in three replicates for each toxin.

2.6. Analysis of Spiked Cereal Samples

The accuracy and precision of the established QB-FLISA were estimated by calculating the recoveries and coefficients of variation of the intra- and inter-assays. Intra-assays were conducted within 1 day in three replicates on each spiked sample. Inter-assays were performed in three replicates on each sample once per day for 3 consecutive days. Cry2A-free cereal samples (corn and rice) were collected from a local farm (Hangzhou, China). The blank cereal sample was spiked with the Cry2A toxin at different concentrations of 200, 1000, and 5000 ng/g. The cereal sample extraction procedure was operated according to the reported method with slight modifications [42]. In brief, 5.0 g of the cereal sample was crushed by a blender. Then, 1.0 g of the cereal sample was weighed and added into 2 mL of extraction buffer (0.2% BSA in 0.1% PBST). After the mixture was gently shaken for 30 min at RT, it was separated by centrifugation (10,000× *g*, 15 min). The supernatant was collected and further diluted 10 times with extraction buffer for QB-FLISA analysis.

3. Results and Discussion

3.1. Preparation and Characterization of the Anti-Cry2A Nanobody

Recombinant expression vector pET-26b(+)-P2 was constructed and confirmed by colony PCR and sequencing. Then, the positive recombinant vector was transformed into *E. coli* Rosetta (DE3). The protein of the nanobody P2 was expressed by IPTG induction and purified on a Ni-NTA affinity chromatography. Subsequently, the expressed and purified nanobody was characterized by SDS-PAGE. Figure 2A shows a target band around 17 kDa observed in the induced *E. coli* cells. The amount of nanobody expressed in the supernatant was obviously higher than that expressed in the inclusion body. The purity of the purified nanobody P2 was over 90%, and the expression yield of the nanobody in the LB culture medium was approximately 30 mg/L.

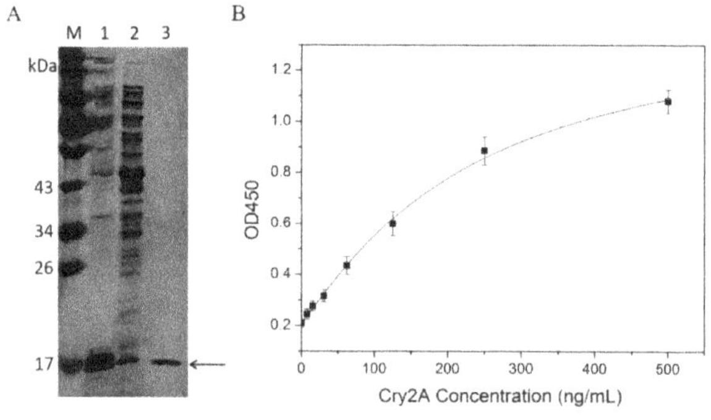

Figure 2. (**A**) SDS-PAGE analysis of the nanobody P2. Lane M: protein marker; Lane 1: the supernatant proteins of the induced *E. coli* cells after sonication; Lane 2: the precipitated protein of the induced *E. coli* cells after sonication; Lane 3: the purified nanobody P2. (**B**) Standard curve (logistic fit) for the Cry2A toxin in the nanobody-based DAS-ELISA. The error bars represent the standard deviation (n = 3).

3.2. Nanobody-Based DAS-ELISA

The purified nanobody P2 was applied to establish a DAS-ELISA for Cry2A toxin analysis. Briefly, 100 μL of the purified nanobody P2 (3.0 μg/mL in PBS) was added into microtiter plates and coated at 4 °C overnight. After blocking was conducted with blocking buffer (3% BSA in 0.1% PBST), serial concentrations of the Cry2A toxins (0–500 ng/mL, 100 μL/well) were added to the microplate and incubated at 37 °C for 1 h. Next, the anti-Cry2A PAb was transferred into the microplate for 1 h of incubation. Then, the plate was washed five times with 0.05% PBST, and 100 μL per well of HRP-labeled goat anti-rabbit antibody was transferred into the plate for another 1 h incubation. Finally, TMB substrates (100 μL/well) were added and incubated for 15 min at 37 °C. The OD450 values were detected by a microtiter plate reader. The results showed that the OD450 value increased with the increase in the concentration of the Cry2A toxin, indicating the good binding activity of the nanobody P2. Meanwhile, a standard curve of the DAS-ELISA was constructed, and the low detection limit (LOD) of the method, calculated as three-times the standard deviation of the blank signal, was 7.83 ng/mL (Figure 2B).

In general, the establishment of the DAS-ELISA depends on conventional MAbs and PAbs. However, the MAb preparation process is complicated and needs a long production period (5–6 months), whereas PAbs cannot achieve the demand of large batches and continual production [43]. Several recognition elements (e.g., nanobodies, peptides, and aptamers), which have emerged in recent years, have been used as substitute for conventional antibodies for the development of immunoassays. Nanobodies can be isolated from a naive phage-displayed nanobody library and further expressed and purified within two weeks, greatly improving the preparation efficiency of antibodies. In addition, nanobodies have excellent characteristics, including high affinity, stability, solubility, and reproducibility, which make them a promising reagent in immunoassays.

3.3. Characterization of QBs-PAb Conjugates

The morphology of the QBs was characterized by TEM. As shown in Figure 3A, the QBs were round and distributed uniformly, with numerous QDs encapsulated in the polymer matrix. The QBs-PAb probe was characterized by fluorescence spectra and DLS analysis. The results showed that the maximum emission wavelength of QBs was 626 nm, while that of the QBs-PAb exhibited a slight blue shift to 622 nm (Figure 3B). In addition, the fluorescence intensity of the QBs-PAb probe slightly decreased compared with that of the QBs, indicating that the PAb modification did not significantly affect the fluorescence emission properties of the QBs. The particle size analysis of the QBs and QBs-PAb is presented in Figure 3C. The average hydrodynamic diameter of the QBs-PAb (154 nm) was larger than the diameter of the QBs (129 nm), indicating that the PAb was coupled on the surface of the QBs. Therefore, these results demonstrated that the QBs-PAb fluorescent probe was successfully synthesized.

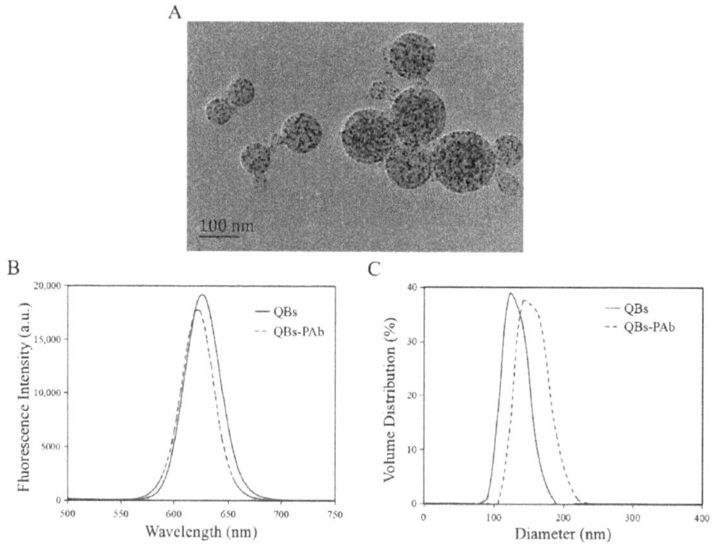

Figure 3. Characterization of QBs and QBs-PAb. (**A**) TEM image of the QBs. (**B**) Fluorescence spectroscopy of QBs and QBs-PAb. (**C**) Hydrodynamic diameter of QBs and QBs-PAb.

3.4. Development of QB-FLISA

A nanobody-based QB-FLISA was established for the determination of Cry2A. The nanobody P2 was used as the capture antibody, and the QB-labeled anti-Cry2A PAb served as the detection probe. Different concentrations of the nanobody P2 (10, 5.0, 2.5, and 1.25 μg/mL) and QBs-PAb (2.0, 1.0, 0.5 and 0.25 μg/mL) were optimized to improve the performance of the proposed QB-FLISA. The LOD value and signal-to-noise ratio were used to evaluate the assay's performance. As shown in Figure 4A, the assay's performance was similar in the P2 concentrations of 10 and 5.0 μg/mL, so the optimal concentration of the nanobody P2 was determined to be 5.0 μg/mL. The fluorescent intensity reduced as the concentration of the QBs-PAb decreased from 2.0 μg/mL to 0.25 μg/mL (Figure 4B). Thus, the concentration of 2.0 μg/mL was selected as the optimal concentration of the QBs-PAb. Subsequently, a nanobody-based QB-FLISA for Cry2A was developed under the optimum experimental conditions. As shown in Figure 5, the developed nanobody-based QB-FLISA exhibited good linearity in the range of 2.6–1000 ng/mL. The LOD value calculated as three-times the standard deviation of the blank signal was 0.41 ng/mL, which was 19-times lower than conventional colorimetric ELISA. This finding indicated that QBs are a favorable fluorescent marker for improving sensitivity in immunoassays for the Cry2A toxin. Comparisons of the proposed QB-FLISA with previously reported methods revealed that the QB-FLISA exhibited a good sensitivity and linearity for Cry2A analysis (Table 1).

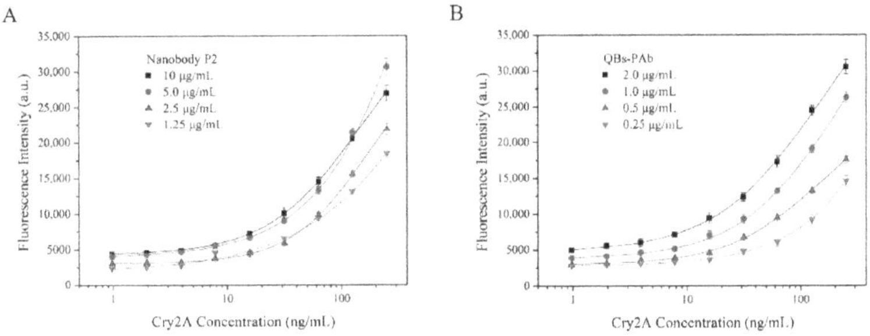

Figure 4. Optimization of nanobody-based QB-FLISA. (**A**) Optimization of the concentrations of the nanobody P2 (10, 5.0, 2.5, and 1.25 μg/mL). (**B**) Optimization of the concentrations of the QBs-PAb (2.0, 1.0, 0.5 and 0.25 μg/mL).

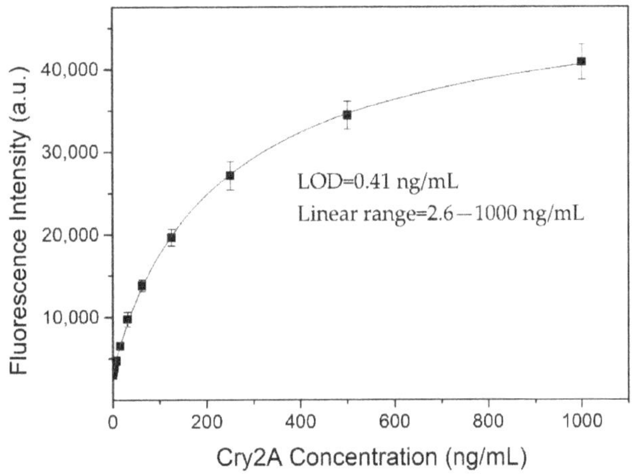

Figure 5. Standard curve (logistic fit) of nanobody-based QB-FLISA for Cry2A toxin analysis under optimized conditions. The error bars represent the standard deviation (n = 3).

Table 1. Comparison of some published methods for Cry toxins' detection.

Detection Method	Antibody	LOD	Working Range	Cry Toxin	Matrix	Reference
Colorimetric ELISA	Traditional antibody	0.27–0.51 ng/mL	0.45–15.71 ng/mL	Cry 1Ie	maize	[2]
Colorimetric IC-ELISA	Human domain antibody	0.029–0.074 μg/mL	0.258–1.407 μg/mL	Cry1Ab, Cry1Ac, Cry1B, Cry1C, and Cry1F	wheat	[8]
Electrochemiluminescent immunosensor	Traditional antibody	3.0 pg/mL	0.010–1.0 ng/mL	Cry1Ab	rice and maize	[10]
Colorimetric ELISA	Traditional antibody	0.47 ng/mL	2.5–100 ng/mL	Cry1Ab	rice	[16]
Colorimetric ELISA	Traditional antibody	15 ng/mL	0.015–32 μg/mL	Cry1	-	[18]
Colorimetric ELISA	ScFv	4.6–9.2 ng/mL	12–250 ng/mL	Cry1Aa, Cry1Ab, and Cry1Ac	-	[19]
Colorimetric ELISA	Phage-displayed peptide	8 ng/mL	10–50.625 ng/mL	Cry2Ad2-3	corn	[20]
QB-FLISA	Nanobody	0.41 ng/mL	2.6–1000 ng/mL	Cry2A	corn and rice	This work

3.5. Cross-Reactivity

The cross-reactivity of the developed QB-FLISA was estimated against other Cry toxins (1000 ng/mL), including Cry1Ab, Cry1B, and Cry3Bb (Figure 6). The results showed a strong fluorescent intensity at 622 nm measured for the Cry2A toxin, but no positive signal was found for the other Cry toxins (Cry1Ab, Cry1B, and Cry3Bb), indicating that the proposed method has no cross-reactivity with the other Cry toxins (1000 ng/mL). These results indicated the good specificity of the developed nanobody-based QB-FLISA for the Cry2A toxin.

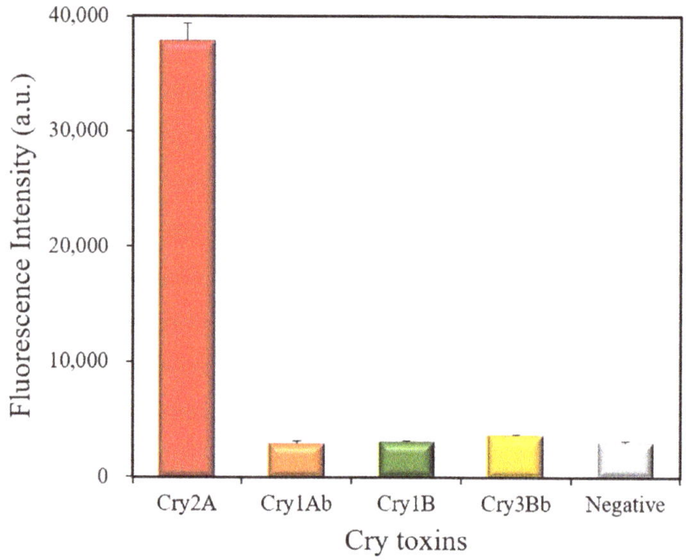

Figure 6. Cross-reactivity of the QB-FLISA with the other Cry toxins (1000 ng/mL).

3.6. Sample Analysis and Validation

Matrix interference is a universal phenomenon in real sample analysis. The dilution of sample extracts is a simple and practical method to eliminate matrix interference. The matrix effect of cereal samples (corn and rice) was estimated at different dilution ratios (1:5, 1:10, 1:20, and 1:40). No evident difference was observed between the 20-fold dilution of extracts and the extraction buffer (Figure 7). Thus, the 20-fold dilution of extracts was selected and used for cereal samples' (corn and rice) analysis.

A spike-and-recovery experiment was conducted to estimate the accuracy and precision of the developed nanobody-based sandwich QB-FLISA. The Cry2A-free cereal samples (corn and rice) spiked with different concentrations of Cry2A toxin (200, 1000, and 5000 ng/g) were tested. As shown in Table 2, the average recoveries of Cry2A in the corn sample ranged from 89.6% to 117.3%, with a coefficient of variation (CV) in the range of 5.5–7.7%. Meanwhile, the recovery in the rice sample ranged from 86.6–109.5%, with a CV in the range of 6.2–8.2%. The Cry2A commercial ELISA kit was used to validate the results of the QB-FLISA. Overall, the results from the two methods correlated well with each other. These results demonstrated that the proposed QB-FLISA is applicable for the detection of Cry2A in cereal samples.

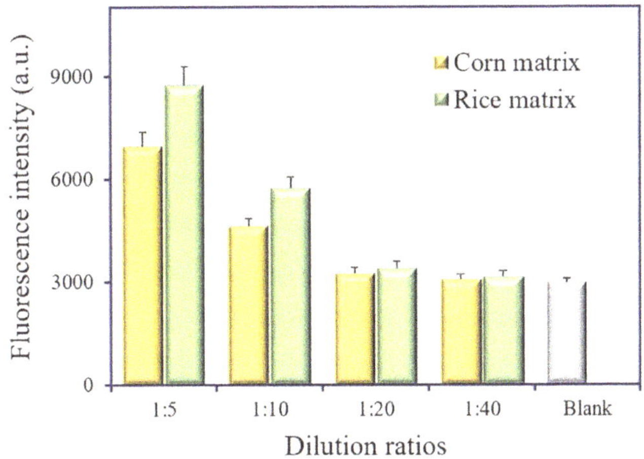

Figure 7. Estimation of the matrix effect in corn and rice samples.

Table 2. Recoveries of Cry2A toxin spiked in cereal samples by nanobody-based QB-FLISA.

Matrix	Spike Level (ng/g)	Intra-Assay (n = 3)			Inter-Assay (n = 3)			Commercial ELISA Kit (n = 3)		
		Mean ± SD (ng/g)	Recovery (%)	CV (%)	Mean ± SD (ng/g)	Recovery (%)	CV (%)	Mean ± SD (ng/g)	Recovery (%)	CV (%)
Corn	200	231.2 ± 16.6	115.6	7.2	223.5 ± 15.7	117.3	7.0	222.8 ± 15.2	111.4	6.8
	1000	962.4 ± 52.8	96.2	5.5	941.1 ± 64.5	94.1	6.9	976.2 ± 57.5	97.6	5.9
	5000	4545 ± 327	90.9	7.2	4480 ± 346	89.6	7.7	4525 ± 371	90.5	8.2
Rice	200	210.8 ± 16.5	105.4	7.8	221.9 ± 18.3	109.5	8.2	233.8 ± 20.4	116.9	8.7
	1000	1027 ± 76.2	102.7	7.4	1018 ± 76.1	101.8	7.5	1082 ± 72.6	108.2	6.7
	5000	4330 ± 269	86.6	6.2	4412 ± 301	88.2	6.8	4635 ± 329	92.7	7.1

4. Conclusions

In this work, an anti-Cry2A nanobody P2 was prepared and applied to develop a DAS-FLISA that was based on QBs for sensitive detection of the Cry2A toxin. The linearity range of the developed nanobody-based QB-FLISA was determined to be 2.6–1000 ng/mL. The LOD of the method was 0.41 ng/mL, which exhibited a 19-fold higher sensitivity compared to the traditional colorimetric ELISA. The spike-and-recovery experiment results showed that the developed method achieved acceptable recoveries (86.6–117.3%) and CVs (5.5–8.2%) for Cry2A detection in cereal samples. These results demonstrated that QBs could be a favorable probe for improving sensitivity in immunoassays. The proposed nanobody-based QB-FLISA may have potential for the sensitive detection of the Cry2A toxin in cereals.

Author Contributions: Conceptualization, methodology, writing–original draft, Y.Q.; methodology, A.Y.; investigation, X.F.; visualization, M.Z.; methodology, H.C.; data curation, B.Z.; validation, W.Q.; project administration, Z.Y.; supervision, writing—review and editing, X.Y. All authors have read and agreed to the published version of the manuscript.

Funding: This research was funded by the National Natural Science Foundation of China (31901781), the Fundamental Research Funds for the Provincial Universities of Zhejiang (2021YW28), the "Pioneer" and "Leading Goose" R&D Program of Zhejiang (2022C02049), the Key planned projects of Zhejiang Province (2021C02058), the Leading Talents in Science and Technology Innovation of Ten Thousand Talents Program in Zhejiang Province (2019R52018), and the Key Research and Development Program of Zhejiang Province (2020C02024).

Institutional Review Board Statement: Not applicable.

Informed Consent Statement: Not applicable.

Data Availability Statement: The data presented in this study are available on request from the corresponding author.

Conflicts of Interest: The authors declare no conflict of interest.

References

1. Bravo, A.; Gill, S.S.; Soberón, M. Mode of action of *Bacillus thuringiensis* Cry and Cyt toxins and their potential for insect control. *Toxicon* **2007**, *49*, 423–435. [CrossRef] [PubMed]
2. Zhang, Y.W.; Zhang, W.; Liu, Y.; Wang, J.H.; Wang, G.Y.; Liu, Y.J. Development of monoclonal antibody-based sensitive ELISA for the determination of Cry1Ie protein in transgenic plant. *Anal. Bioanal. Chem.* **2016**, *408*, 8231–8239. [CrossRef] [PubMed]
3. Bravo, A.; Gómez, I.; Porta, H.; García-Gómez, B.I.; Rodriguez-Almazan, C.; Pardo, L.; Soberón, M. Evolution of *Bacillus thuringiensis* Cry toxins insecticidal activity. *Microb. Biotechnol.* **2013**, *6*, 17–26. [CrossRef] [PubMed]
4. Lu, K.; Gu, Y.; Liu, X.; Lin, Y.; Yu, X.Q. Possible insecticidal mechanisms mediated by immune response related Cry-binding proteins in the midgut juice of *Plutella xylostella* and *Spodoptera exigua*. *J. Agric. Food Chem.* **2017**, *65*, 2048–2055. [CrossRef]
5. Pardo-López, L.; Soberón, M.; Bravo, A. *Bacillus thuringiensis* insecticidal three-domain Cry toxins: Mode of action, insect resistance and consequences for crop protection. *FEMS Microbiol. Rev.* **2013**, *37*, 3–22. [CrossRef]
6. Lambert, B.; Pefereon, M. Insecticidal promise of *Bacillus thuringiensis*. *BioScience* **1992**, *42*, 112–122. [CrossRef]
7. de Maagd, R.A.; Bosch, D.; Stiekema, W. *Bacillus thuringiensis* toxin-mediated insect resistance in plants. *Trends Plant Sci.* **1999**, *4*, 9–13. [CrossRef]
8. Xu, C.; Zhang, X.; Liu, X.; Liu, Y.; Hu, X.; Zhong, J.; Zhang, C.; Liu, X. Selection and application of broad-specificity human domain antibody for simultaneous detection of Bt Cry toxins. *Anal. Biochem.* **2016**, *512*, 70–77. [CrossRef]
9. Lu, X.; Jiang, D.; Yan, J.; Ma, Z.; Luo, X.; Wei, T.; Xu, Y.; He, Q. An ultrasensitive electrochemical immunosensor for Cry1Ab based on phage displayed peptides. *Talanta* **2018**, *179*, 646–651. [CrossRef]
10. Gao, H.; Wen, L.; Wu, Y.; Fu, Z.; Wu, G. An ultrasensitive label-free electrochemiluminescent immunosensor for measuring Cry1Ab level and genetically modified crops content. *Biosens. Bioelectron.* **2017**, *97*, 122–127. [CrossRef]
11. Faheem, A.; Qin, Y.; Nan, W.; Hu, Y. Advances in the immunoassays for detection of *Bacillus thuringiensis* crystalline toxins. *J. Agric. Food Chem.* **2021**, *69*, 10407–10418. [CrossRef] [PubMed]
12. Kamle, S.; Ali, S. Genetically modified crops: Detection strategies and biosafety issues. *Gene* **2013**, *522*, 123–132. [CrossRef] [PubMed]
13. Albright, V.C., III; Hellmich, R.L.; Coats, J.R. A review of cry protein detection with enzyme-linked immunosorbent assays. *J. Agric. Food Chem.* **2016**, *64*, 2175–2189. [CrossRef] [PubMed]
14. Fraiture, M.A.; Herman, P.; Taverniers, I.; De Loose, M.; Deforce, D.; Roosens, N.H. Current and new approaches in GMO detection: Challenges and solutions. *Biomed Res. Int.* **2015**, *2015*, 392872. [CrossRef] [PubMed]
15. Freitas, M.; Correr, W.; Cancino-Bernardi, J.; Barroso, M.F.; Delerue-Matos, C.; Zucolotto, V. Impedimetric immunosensors for the detection of Cry1Ab protein from genetically modified maize seeds. *Sens. Actuator B Chem.* **2016**, *237*, 702–709. [CrossRef]
16. Dong, S.; Zhang, X.; Liu, Y.; Zhang, C.; Xie, Y.; Zhong, J.; Xu, C.; Liu, X. Establishment of a sandwich enzyme-linked immunosorbent assay for specific detection of *Bacillus thuringiensis* (Bt) Cry1Ab toxin utilizing a monoclonal antibody produced with a novel hapten designed with molecular model. *Anal. Bioanal. Chem.* **2017**, *409*, 1985–1994. [CrossRef] [PubMed]
17. Kamle, S.; Ojha, A.; Kumar, A. Development of an enzyme linked immunosorbant assay for the detection of Cry2Ab protein in transgenic plants. *GM Crops* **2011**, *2*, 118–125. [CrossRef]
18. Dong, S.; Zhang, C.; Zhang, X.; Liu, Y.; Zhong, J.; Xie, Y.; Xu, C.; Ding, Y.; Zhang, L.; Liu, X. Production and characterization of monoclonal antibody broadly recognizing Cry1 toxins by use of designed polypeptide as hapten. *Anal. Chem.* **2016**, *88*, 7023–7032. [CrossRef]
19. Dong, S.; Gao, M.; Bo, Z.; Guan, L.; Hu, X.; Zhang, H.; Liu, B.; Li, P.; He, K.; Liu, X.; et al. Production and characterization of a single-chain variable fragment antibody from a site-saturation mutagenesis library derived from the anti-Cry1A monoclonal antibody. *Int. J. Biol. Macromol.* **2020**, *149*, 60–69. [CrossRef]
20. Wang, Y.; Zhang, X.; Xie, Y.; Wu, A.; Zai, X.; Liu, X. High-affinity phage-displayed peptide as a recognition probe for the detection of Cry2Ad2-3. *Int. J. Biol. Macromol.* **2019**, *137*, 562–567. [CrossRef]
21. Peyman, B.; Mozafar, M.; Ali, H. ScFv improvement approaches. *Protein Pept. Lett.* **2018**, *25*, 222–229. [CrossRef]
22. Liscano, Y.; Oñate-Garzón, J.; Delgado, J.P. Peptides with Dual Antimicrobial–Anticancer Activity: Strategies to Overcome Peptide Limitations and Rational Design of Anticancer Peptides. *Molecules* **2020**, *25*, 4245. [CrossRef] [PubMed]
23. Hamers-Casterman, C.; Atarhouch, T.; Muyldermans, S.; Robinson, G.; Hammers, C.; Songa, E.B.; Bendahman, N.; Hammers, R. Naturally occurring antibodies devoid of light chains. *Nature* **1993**, *363*, 446–448. [CrossRef] [PubMed]
24. Salvador, J.P.; Vilaplana, L.; Marco, M.P. Nanobody: Outstanding features for diagnostic and therapeutic applications. *Anal. Bioanal. Chem.* **2019**, *411*, 1703–1713. [CrossRef] [PubMed]
25. Hu, Y.; Zhang, C.; Yang, F.; Lin, J.; Wang, Y.; Wu, S.; Sun, Y.; Zhang, B.; Lv, H.; Ji, X.; et al. Selection of Specific Nanobodies against Lupine Allergen Lup an 1 for Immunoassay Development. *Foods* **2021**, *10*, 2428. [CrossRef]
26. Yang, J.; Si, R.; Wu, G.; Wang, Y.; Fang, R.; Liu, F.; Wang, F.; Lei, H.; Shen, Y.; Zhang, Q.; et al. Preparation of Specific Nanobodies and Their Application in the Rapid Detection of Nodularin-R in Water Samples. *Foods* **2021**, *10*, 2758. [CrossRef]

27. Muyldermans, S.; Baral, T.N.; Cortez Retamozzo, V.; De Baetselier, P.; De Genst, E.; Kinne, J.; Leonhardt, H.; Magez, S.; Nguyen, V.K.; Revets, H.; et al. Camelid immunoglobulins and nanobody technology. *Vet. Immunol. Immunop.* **2009**, *128*, 178–183. [CrossRef]
28. Kumar, B.; Malhotra, K.; Fuku, R.; Van Houten, J.; Qu, G.Y.; Piunno, P.A.E.; Krull, U.J. Recent trends in the developments of analytical probes based on lanthanide-doped upconversion nanoparticles. *TrAC Trends Anal. Chem.* **2021**, *139*, 116256. [CrossRef]
29. Zhao, Y.; Zhang, P.; Wang, J.; Zhou, L.; Yang, R. A novel electro-driven immunochromatography assay based on upconversion nanoparticles for rapid pathogen detection. *Biosens. Bioelectron.* **2020**, *152*, 112037. [CrossRef] [PubMed]
30. Wang, X.; Wu, X.; Lu, Z.; Tao, X. Comparative Study of Time-Resolved Fluorescent Nanobeads, Quantum Dot Nanobeads and Quantum Dots as Labels in Fluorescence Immunochromatography for Detection of Aflatoxin B_1 in Grains. *Biomolecules* **2020**, *10*, 575. [CrossRef]
31. Chen, B.; Shen, X.; Li, Z.; Wang, J.; Li, X.; Xu, Z.; Shen, Y.; Lei, Y.; Huang, X.; Wang, X.; et al. Antibody Generation and Rapid Immunochromatography Using Time-Resolved Fluorescence Microspheres for Propiconazole: Fungicide Abused as Growth Regulator in Vegetable. *Foods* **2022**, *11*, 324. [CrossRef] [PubMed]
32. Xiang, X.; Ye, Q.; Sheng, Y.; Li, F.; Zhou, B.; Shao, Y.; Wang, C.; Zhang, J.; Xue, L.; Chen, M.; et al. Quantitative detection of aflatoxin B_1 using quantum dots-based immunoassay in a recyclable gravity-driven microfluidic chip. *Biosens. Bioelectron.* **2021**, *190*, 113394. [CrossRef] [PubMed]
33. Tong, W.; Fang, H.; Xiong, H.; Wei, D.; Leng, Y.; Hu, X.; Huang, X.; Xiong, Y. Eco-Friendly Fluorescent ELISA Based on Bifunctional Phage for Ultrasensitive Detection of Ochratoxin A in Corn. *Foods* **2021**, *10*, 2429. [CrossRef] [PubMed]
34. Zhao, L.; Jin, J.; Zhu, W.; Zuo, Y.; Song, Y. Detection of Pyrethroids in Food by Immunofluorescence Enhanced Method Based on Three-Layer Core-Shell Structure Upconversion Materials. *Foods* **2022**, *11*, 990. [CrossRef]
35. Ouyang, Q.; Wang, L.; Ahmad, W.; Rong, Y.; Li, H.; Hu, Y.; Chen, Q. A highly sensitive detection of carbendazim pesticide in food based on the upconversion-MnO_2 luminescent resonance energy transfer biosensor. *Food Chem.* **2021**, *349*, 129157. [CrossRef] [PubMed]
36. Hou, S.; Ma, J.; Cheng, Y.; Wang, H.; Sun, J.; Yan, Y. Quantum dot nanobead-based fluorescent immunochromatographic assay for simultaneous quantitative detection of fumonisin B_1, dexyonivalenol, and zearalenone in grains. *Food Control* **2020**, *117*, 107331. [CrossRef]
37. Liu, B.; Li, P.; Wang, Y.; Guo, Y.; Zhang, H.; Dong, S.; Xiong, Y.; Zhang, C. Quantum dot submicrobead–based immunochromatographic assay for the determination of parathion in agricultural products. *Food Anal. Method.* **2020**, *13*, 1736–1745. [CrossRef]
38. Li, P.; Yang, C.; Liu, B.; Wu, Q.; Wang, Y.; Dong, S.; Zhang, H.; Vasylieva, N.; Hammock, B.D.; Zhang, C. Sensitive Immunochromatographic Assay Using Highly Luminescent Quantum Dot Nanobeads as Tracer for the Detection of Cyproheptadine Hydrochloride in Animal-Derived Food. *Front. Chem.* **2020**, *8*, 575. [CrossRef]
39. Qie, Z.; Yan, W.; Gao, Z.; Meng, W.; Xiao, R.; Wang, S. Ovalbumin antibody-based fluorometric immunochromatographic lateral flow assay using CdSe/ZnS quantum dot beads as label for determination of T-2 toxin. *Microchim. Acta* **2019**, *186*, 816. [CrossRef]
40. Qiu, Y.; Li, P.; Liu, B.; Liu, Y.; Wang, Y.; Tao, T.; Xu, J.; Hammock, B.D.; Liu, X.; Guan, R.; et al. Phage-displayed nanobody based double antibody sandwich chemiluminescent immunoassay for the detection of Cry2A toxin in cereals. *Food Agric. Immunol.* **2019**, *30*, 924–936. [CrossRef]
41. Shao, Y.; Duan, H.; Guo, L.; Leng, Y.; Lai, W.; Xiong, Y. Quantum dot nanobead-based multiplexed immunochromatographic assay for simultaneous detection of aflatoxin B_1 and zearalenone. *Anal. Chim. Acta* **2018**, *1025*, 163–171. [CrossRef] [PubMed]
42. Xu, C.; Zhang, C.; Zhong, J.; Hu, H.; Luo, S.; Liu, X.; Zhang, X.; Liu, Y.; Liu, X. Construction of an Immunized Rabbit Phage Display Library for Selecting High Activity against *Bacillus thuringiensis* Cry1F Toxin Single-Chain Antibodies. *J. Agric. Food Chem.* **2017**, *65*, 6016–6022. [CrossRef] [PubMed]
43. Sun, Z.; Duan, Z.; Liu, X.; Deng, X.; Tang, Z. Development of a Nanobody-Based Competitive Dot ELISA for Visual Screening of Ochratoxin A in Cereals. *Food Anal. Mehtods* **2017**, *10*, 3558–3564. [CrossRef]

foods

MDPI

Article

Identification and Application of Two Promising Peptide Ligands for the Immunodetection of Imidacloprid Residue

Tianyang You [1,2,†], Yuan Ding [1,2,†], Yue Huang [1,2], Yang Lu [1,2], Minghua Wang [1,2] and Xiude Hua [1,2,*]

1 College of Plant Protection, Nanjing Agricultural University, Nanjing 210095, China
2 State & Local Joint Engineering Research Center of Green Pesticide Invention and Application, Nanjing 210095, China
* Correspondence: huaxiude@njau.edu.cn; Tel./Fax: +86-25-84395479
† These authors contributed equally to this work.

Abstract: As the most widely used neonicotinoid insecticide, it is of great significance to explore the immunoreagents and immunoassays for imidacloprid (IMI) residue. In immunoassays, specific peptide ligands, such as peptidomimetic and anti-immunocomplex peptides, are regarded as promising substitutes for chemical haptens. In the present work, we identified thirty sequences of peptidomimetics and two sequences of anti-immunocomplex peptides for IMI from three phage pVIII display cyclic peptide libraries, in which the anti-immunocomplex peptides are the first reported noncompetitive reagents for IMI. The peptidomimetic 1-9-H and anti-immunocomplex peptide 2-1-H that showed the best sensitivity were utilized to develop competitive and noncompetitive phage enzyme-linked immunosorbent assays (P-ELISAs), with a half inhibition concentration of 0.55 ng/mL for competitive P-ELISA and a half-saturation concentration of 0.35 ng/mL for noncompetitive P-ELISA. The anti-immunocomplex peptide was demonstrated to greatly improve the specificity compared with competitive P-ELISA. In addition, the accuracy of proposed P-ELISAs was confirmed by recovery analysis and HPLC verification in agricultural and environmental samples. These results show that the peptide ligands identified from phage display library can replace chemical haptens in the immunoassays of IMI with the satisfactory performance.

Keywords: imidacloprid; anti-immunocomplex peptide; peptidomimetic; noncompetitive immunoassay; competitive immunoassay

Citation: You, T.; Ding, Y.; Huang, Y.; Lu, Y.; Wang, M.; Hua, X. Identification and Application of Two Promising Peptide Ligands for the Immunodetection of Imidacloprid Residue. *Foods* **2022**, *11*, 3163. https://doi.org/10.3390/foods11203163

Academic Editor: Simon Haughey

Received: 15 August 2022
Accepted: 8 October 2022
Published: 11 October 2022

Publisher's Note: MDPI stays neutral with regard to jurisdictional claims in published maps and institutional affiliations.

Copyright: © 2022 by the authors. Licensee MDPI, Basel, Switzerland. This article is an open access article distributed under the terms and conditions of the Creative Commons Attribution (CC BY) license (https://creativecommons.org/licenses/by/4.0/).

1. Introduction

Immunoassay plays an increasingly important role in pesticide residue detection. Although the precision and reliability of chromatography-based methods are apparent, the advantages of immunoassay, such as its low-cost, simple operation and portable detection device, make it the preferable approach for screening large samples and for on-site detection. The establishment of immunoassay is realized by the recognition between immunorecognition reagents (antibody and antigen) [1,2]. The chemical hapten and monoclonal antibody (mAb) are the classic combination of immunorecognition reagents, which has achieved great success in the immunoassays for pesticide residue [3,4]. Even though there is still no clear design rule for high quality hapten, trial and error are the commonly used methods [5–7]. Given the great variety of pesticides, preparation of high quality hapten for each pesticide means a tremendous amount of work. Besides, the competitive immunoassay based on chemical hapten generates a negative correlation signal that makes the detection results not intuitive.

In recent years, the specific peptide screened from the phage display peptide library has shown the potential to replace chemical hapten as the next-generation immunoreagent in pesticide residue immunoassay [8]. The peptides can work in two different ways. One way is the same as with chemical haptens, which directly reacts with the antibody (called peptidomimetic) [9]. The other way is more special, in which the peptide reacts with the immunocomplex between antibody and antigen (called as anti-immunocomplex peptide) to establish the noncompetitive immunoassay [10]. The noncompetitive format provides a positive correlation signal that makes the detection results more friendly. Unlike open sandwich immunoassays or the noncompetitive immunoassay based on an anti-idiotype antibody, anti-immunocomplex peptide is easier to obtain so that it is increasingly being reported in pesticide residue immunoassay [11–14]. These two types of peptides generally improved the sensitivity or specificity of immunoassays in previous reports. More importantly, the screening procedure can be completed within a week in a basic biology laboratory, which is of great significance to preparing the immunoreagent of various pesticides.

Imidacloprid (IMI) is the first registered and most widely used neonicotinoid. It acts on the nicotinic acetylcholine receptors of pests so that it disturbs the normal nerve conduction, it is mainly used against Homoptera, Thysanoptera, Coleoptera, Diptera and Lepidoptera pests. Although IMI has been completely banned from outdoor use by the European Commission (EU) to protect honeybees, it is still registered and used globally in crop and ornamentals protection, urban pest control, veterinary applications, and fish farming, owing to its excellent systemic and contact activity. The widespread use and persistence property render IMI detectable in all kinds of agricultural, water, and soil samples worldwide [15–17]. Besides the high toxicity to honeybees, there is some research which showed that IMI exposure was positively correlated with several diseases in newborns [18–20] and oxidative DNA damage in adults [21–23]. Governments regulate the maximum residue limits (MRLs) of imidacloprid in various food. China has regulated the MRLs of imidacloprid in 123 kinds of foods, which range from 0.01 mg kg^{-1} to 10 mg kg^{-1} (GB2763-2021). The EU has regulated the MRLs of imidacloprid in 381 kinds of food with the range of 0.01 mg kg^{-1} to 15 mg kg^{-1} (EU 2021/1881). The U.S. Environmental Protection Agency has regulated the MRLs of imidacloprid in 134 kinds of food with the range of 0.02 mg kg^{-1} to 240 mg kg^{-1}. Even though IMI has been banned from outdoor use in EU, it is still necessary to monitor IMIs concentration in the environment. So, it is desirable to develop the analytical method for imidacloprid residue. Currently, dozens of immunoassays for IMI residue have been reported, most of which are competitive immunoassays based on chemical haptens [24–29], but there is no reported noncompetitive immunoreagent for IMI.

In our recent study, we optimized the pVIII phage display peptide system to improve the panning success rate [13]. In the present work, IMI was determined as a target and an anti-IMI mAb 3D11B12E5 was employed to screen above the pVIII phage display peptide libraries. Then, peptidomimetic and anti-immunocomplex peptides were used to develop competitive and noncompetitive phage enzyme-linked immunosorbent assays (P-ELISAs) for IMI, and then the sensitivities and specificities of the P-ELISAs were compared. Finally, the competitive and noncompetitive P-ELISAs were utilized to test the IMI spiked samples, which the results were verified with high performance liquid chromatography (HPLC).

2. Materials and Methods

2.1. Reagents

Imidacloprid (99.5%) and imidaclothiz (97.8%) were obtained from Dr. Ehrenstorfer (Augsburg, Germany) and from Jiangshan Agrochemical and Chemicals Co., Ltd. (Nantong, China), respectively. Other compounds for specificity test were all purchased from Dr. Ehrenstorfer (Augsburg, Germany). The ER 2738 and helper phage M13KO7 were purchased from New England Biolabs (Ipswich, UK). Tween-20 was purchased from Solarbio (Beijing, China). Skimmed milk was purchased from Bio-Rad (Hercules, CA, USA). The HRP labelled anti-M13 antibody was purchased from SinoBiological (Beijing, China). The ninety-six-well microplates were purchased from Corning Costar (Corning, NY, USA). The mAb 3D11B12E5 purified by protein A column and phage-displayed cyclic 8, 9, 10-residue random peptide libraries were prepared and stored in our own laboratory [13,30]. The sequencing primer was synthesized by Genscript Bio. (Nanjing, China).

2.2. Biopanning

Anti-IMI mAb 3D11B12E5 diluted in PBS was added to nine wells of microplate at 4 °C overnight (100 μL/well). The plate was washed three times by PBST (PBS containing tween-20) and blocked with 300 μL 5% skimmed milk (dissolved in PBS) at 37 °C for 2 h. For peptidomimetic panning, 100 μL/well mixture of phage-displayed cyclic 8-, 9-, 10-residue random peptide libraries was diluted using 5% skimmed milk, and added to three coated wells, and incubated at room temperature for 1 h. The plate was washed ten times with PBST, incubated with PBS for 30 min, and then washed ten times with PBS. After washing, 100 μL/well IMI standard solution was added and incubated at room temperature for 1 h to competitively elute the bound phage. For immunocomplex panning, 100 μL/well IMI standard solution (10 μg/mL) was firstly injected to form immunocomplex for 1 h. After discarding the excess IMI, the mixture of phage-displayed peptide libraries used in peptidomimetic panning was added and incubated at room temperature for another 1 h. After washing, the bound phage was eluted by an elution buffer (0.2 M Gly-HCl containing 1 mg/mL BSA, pH 2.2) for 15 min, and immediately neutralized to pH 7.4 using 1 M Tris-HCl (pH 9.1). The eluted phages were titer-determined and amplified [13] for subsequent panning.

A total of three rounds of panning were performed under different panning conditions. The concentrations of PBST used in the first, second, and third rounds were 0.1%, 0.3%, and 0.5%. The coating concentrations of mAb were 10, 5, and 2.5 μg/mL for the first, second, and third rounds. The quantities of phage inputted in the first, second, and third rounds of panning were 3×10^{11}, 3×10^{10} and 3×10^{10} pfu. The concentrations of IMI standard solution in peptidomimetic panning were 10, 5 and 2.5 μg/mL. After three rounds of panning, a total of 32 clones from the last titer-determining plate were picked and amplified for the identification of positive clones using P-ELISAs.

2.3. P-ELISAs

100 μL/well mAb 3D11B12E5 diluted in PBS was coated in the microplate at 4 °C overnight. After being washed three times with 0.5‰ P BST, the plate was blocked with 300 μL/well 5% skimmed milk (diluted in PBS) at 37 °C for 1.5 h. After washing, fifty microliters of IMI standard or sample solution and fifty microliters of phage-displayed peptide diluted using 5% skimmed milk was added to the plate, and reacted at 37 °C for 1 h. After washing, 100 μL anti-M13 antibody (HRP) diluted with 0.05% PBST (0.2 μg/mL, containing 0.5% skimmed milk) was added to the recognized phage particles. After washing, 100 μL of peroxidase substrate (10 mL 0.1 M citric acid and dibasic sodium phosphate buffer (pH 5.5), 32 μL 0.75% H_2O_2, and 100 μL 10 mg/mL TMB in dimethyl sulfoxide) was added and reacted at 37 °C for 15 min, then stopped with 50 μL 2 M H_2SO_4. The optical density at 450 nm (OD_{450}) was determined with a SpectraMax M5 microplate reader. The dose-response curve was fitted by logistic equation using the IMI concentration on the X axis and OD_{450} value on the Y axis.

2.4. Optimization of P-ELISAs

The optimal concentrations of mAb 3D11B12E5 and phage-displayed peptides were determined by the checkboard method, in which a series of concentrations of mAb 3D11B12E5 were combined with various quantities of phage-displayed peptides to run P-ELISAs, respectively. For determining the optimal buffer, the dose-response curves of P-ELISAs were created in various PBS buffers, which contained different Na^+ concentrations (from 0.035 to 2.4 mol/L), pH values (from 5.0 to 9.0), and methanol contents (from 1.25% to 10%).

2.5. Specificity

The specificities of P-ELISAs, defined as the cross-reactivities (CRs) with other analogs of IMI, were calculated as follows: CR = [IC_{50} (or SC_{50}) $_{IMI}$/IC_{50} (or SC_{50}) $_{analogs}$] × 100%.

2.6. Analysis of Spiked Samples

The accuracies of proposed P-ELISAs for IMI detection in various agricultural (wheat, brown rice, potato, cabbage, pear, and orange) and environmental samples (paddy water and soil) were evaluated. The samples were collected from Nanjing, China. The agricultural samples were chopped and homogenized. The paddy water samples were filtered, while the soil samples were air-dried, crushed, and passed through a 2 mm sieve. All of the samples were stored at −20 °C. The samples were confirmed not to be contaminated with IMI by HPLC.

The samples were spiked with various concentrations of IMI standard solutions and stood overnight. Before the analysis of the spiked samples, their matrix effect on P-ELISAs were firstly investigated by preparing the dose-response curves in series diluted blank extracts, then compared with the PBS-based standard curve to determine the dilution factors. After dilution, the paddy water was directly detected without further processing. For other solid samples, 5 g samples were added to a 50 mL centrifuge tube, then extracted by 10 mL optimized PBS containing 45% methanol. The tube was vortexed at 300 rpm for 5 min, sonicated for 15 min, then vortexed for another 5 min. After centrifuging at 4000 rpm for 5 min, the extracts were collected, and the extracts of cabbage, pear, potato, and orange were adjusted to 15 mL by PBS. The extracts of spiked samples were tested by P-ELISAs after proper dilutions.

2.7. Verification of P-ELISAs by HPLC

IMI standard solutions with unknown concentrations were spiked into pear and soil samples, respectively. For P-ELISAs, the samples were processed and analyzed as above. For HPLC, the samples were processed and analyzed as previously described [9].

3. Results

3.1. Panning Results

The peptide libraries used in this study are displayed on phage pVIII in high-density, whose expressions are controlled by the *lac* promoter/operator. Such a display strategy may increase the apparent affinity and weaken the growth bias among different phages, thus serving as a footstone for the blended panning strategy using mixed libraries [13].

The phage titer of each panning is shown in Table S1. While the screening condition got more stringent, the titers showed an increase of 100–1000 fold after the third round of panning, indicating an enrichment of desired peptidomimetics or anti-immunocomplex peptides. After identification, thirty clones showed decreased signals in the presence of IMI (Figure 1a), and twenty eight clones showed enhanced signals in the presence of IMI (Figure 1b). The sequences of these positive clones are shown in Table 1, in which a total of thirty sequences of peptidomimetics and two sequences of anti-immunocomplex peptides are identified. Most sequences of peptidomimetics contained the consensus motif TPAG, while there is no consensus motif among the anti-immunocomplex peptides. While the 8-, 9-, 10-residue peptide libraries were mixed equally for panning, peptides from the 9-residue peptide library were not included in the anti-immunocomplex peptides. It's a common fact that peptides bound to the immunocomplex of a given target may not exist in one particular library. Panning through multiple libraries is necessary to increase the chance, and a blended manner will help reduce the heavy workload.

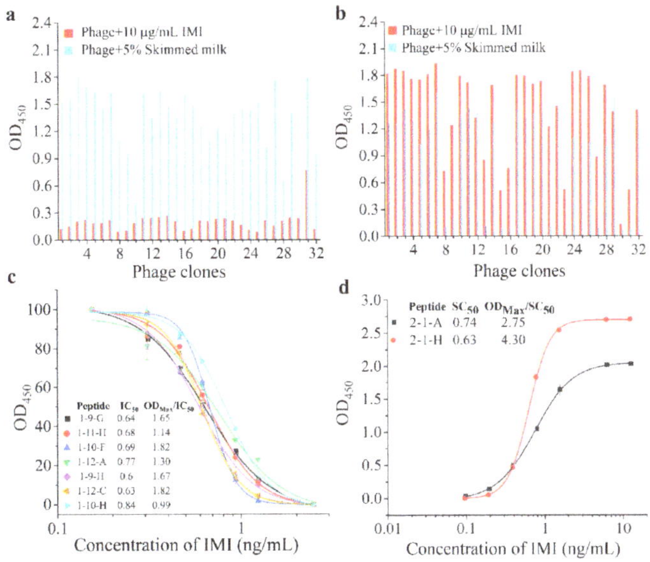

Figure 1. Panning results. (**a**) The screen of peptidomimetics. (**b**) The screen of anti-immunocomplex peptides. (**c**) Dose-response curve of different peptidomimetics. (**d**) Dose-response curve of different anti-immunocomplex peptides.

Table 1. Sequence of isolated peptidomimetics and anti-immunocomplex peptides.

Clone	Amino Acid Quantity	Amino Acid Sequence	Clone	Amino Acid Quantity	Amino Acid Sequence
		Peptidomimetics (consensus TPAG)			
1-9-A	8	CFETPAGLMC	1-11-A	8	CTMPDPAGMC
1-9-B	8	CEMTPAGWLC	1-11-B	8	CLETPAGLAC
1-9-C	10	MSPAGIWDAQC	1-11-C	8	CVASPAGLVC
1-9-D	8	CESTPAGYFC	1-11-D	8	CAATPAGLVC
1-9-E	9	CTDPAGMLASC	1-11-E	8	CHSSPAGFVC
1-9-F	9	CTFTPSVRRIC	1-11-F	8	CVETPAGFYC
1-9-G	8	CSMSPAGPIC	1-11-G	8	CEWTPAGWVC
1-9-H	8	CVPTPAGDFC	1-11-H	8	CVDTPAGLYC
1-10-A	8	CVSTPAGLTC	1-12-A	8	CPLTPAGPVC
1-10-C	8	CVMSPAGPVC	1-12-B	8	CEMTPAGLAC
1-10-D	8	CVSSPGGLVC	1-12-C	8	CEMTPAGPVC
1-10-E	8	CEQTPAGLVC	1-12-D	9	CEMNPAGIRIC
1-10-F	8	CEYSPAGVIC	1-12-E	8	CEDSPAGWIC
1-10-G	8	CLMTPAGPSC	1-12-G	8	CHESPGGMIC
1-10-H	8	CEQTPAGLMC	1-12-H	8	CTMSPGGWIC
		Anti-immunocomplex peptides (No consensus sequence)			
2-1-A	8	CWCIEDCSNC (18) [1]	2-1-H	10	CVWDGDVGIMYC (9) [1]

[1] Quantity of same sequence clones.

3.2. Optimization of P-ELISAs

The sensitivities of seven peptidomimetics (1-9-G, 1-9-H, 1-10-F, 1-10-H, 1-11-H, 1-12-A, 1-12-C) that showed enhanced signal differences in the presence of IMI, and all anti-immunocomplex peptides (2-1-A, 2-1-H), were evaluated by P-ELISAs. The optimal concentration of mAb and input quantities of phage-displayed peptides were firstly determined. The dosages of mAb 3D11B12E5 and phage-displayed peptides that generated the

$OD_{450} \approx 1$ were chosen to develop dose-response curve for competitive P-ELISA (Table S2 and Figure 1c). The dosages that produced the highest S/N were chosen to develop dose-response curves for noncompetitive P-ELISA (Figure S1 and Figure 1d). The peptidomimetics 1-9-H showed the lowest IC_{50} and the highest maximum OD_{450} (OD_{max})/IC_{50} were selected for subsequent experiments, while the anti-immunocomplex peptides 2-1-H were selected.

The binding between antigen and antibody is mediated by Van der Waals, hydrogen and ion force, which will be influenced by the pH and ionic strength in reaction environment [31]. As an effective solvent for pesticide extraction, methanol is also inevitable in their immunoassays. To achieve the best sensitivity, these conditions in P-ELISAs were subsequently optimized. The conditions that generate the dose-response curve with the lowest IC_{50} (or SC_{50}) and the highest OD_{max}/IC_{50} (or OD_{max}/SC_{50}) were considered as the optimums. As shown in Figure 2, the PBS buffer containing 0.14 M Na^+ and pH 6.0 was chosen to develop competitive P-ELISA, and the PBS buffer containing 0.07 M Na^+ and pH 8.0 was chosen to develop noncompetitive P-ELISA. Reaction between the antibody and phage display peptides was significantly disturbed when the methanol concentrations in the buffer were higher than 2.5% (Figure 2e,f), so the maximum tolerance of P-ELISAs to methanol was 2.5%. In order to avoid the interference caused by different methanol concentrations in the pretreatment of various samples, the methanol concentrations were all determined as 2.5%.

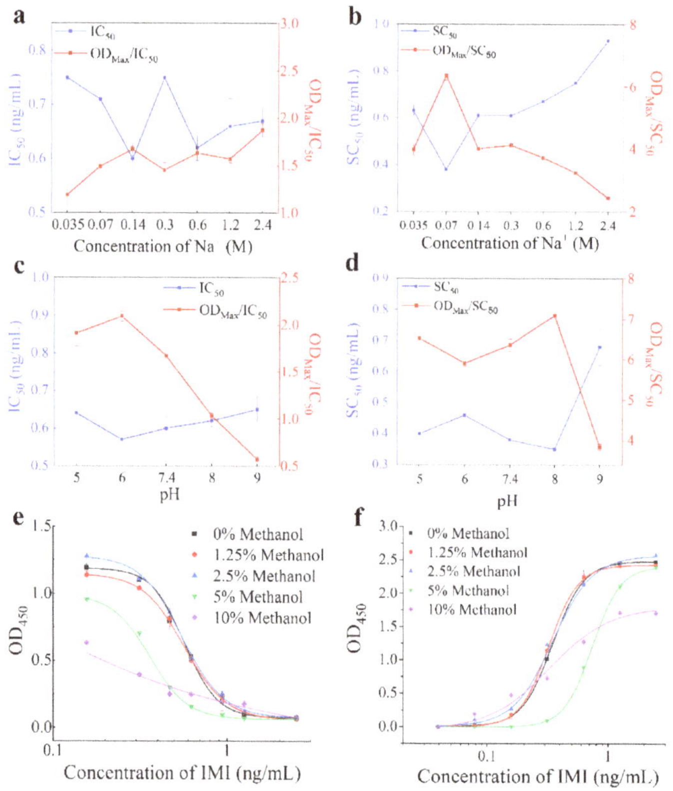

Figure 2. Optimization of P-ELISAs. (**a**,**c**,**e**) The concentration of Na^+, pH and methanol for competitive P-ELISA. (**b**,**d**,**f**) The concentration of Na^+, pH and methanol for noncompetitive P-ELISA.

3.3. Sensitivity of P-ELISAs

The dose-response curves of two P-ELISAs are shown in Figure 3. The IC_{50} of competitive P-ELISA was 0.55 ng/mL, with the limit of detection (LOD, IC_{10}) of 0.30 ng/mL and detection range (IC_{10}–IC_{90}) of 0.30–1.00 ng/mL, while the SC_{50} of noncompetitive P-ELISA was 0.35 ng/mL, with the LOD of 0.15 ng/mL and detection range (SC_{10}–SC_{90}) of 0.15–0.80 ng/mL. The noncompetitive P-ELISA showed slightly improved sensitivity than competitive P-ELISA. We have summarized the reported sensitivities of the basic immunodetection, that is ELISA, for the detection of IMI (Table S3). Except for the P-ELISAs reported here, there are two other competitive P-ELISAs using phage displayed peptidomimetics [9,32] and six competitive ELISAs using chemical haptens [24–29]. The proposed P-ELISAs showed better sensitivities than most reported ELISAs, just lower than another competitive P-ELISA using a pIII displayed linear 12-amino acid peptidomimetic. The latter immunoassay, which was reported by Du et al. [9], achieves the highest sensitivity (IC_{50} = 0.067 ng/mL). The HRP labelled anti-M13 antibody used in this work is targeted at phage pVIII, whose binding may be hampered by the phage pVIII displayed peptides, thus causing a lower signal amplification effect and leading to the poor sensitivity of P-ELISAs. It is worth mentioning that isolated peptides here are cyclic peptides, which may mean better stability. Moreover, the noncompetitive immunoreagent for imidacloprid is reported for the first time, which can derive a series of noncompetitive immunoassays for imidacloprid with more intuitive detection results. It has advantages in the development of visual detection.

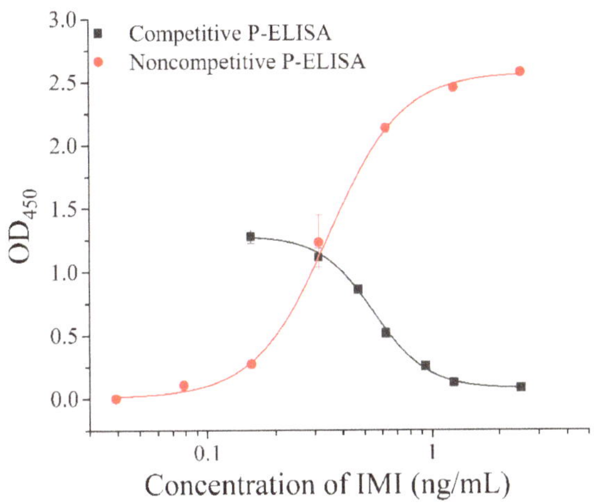

Figure 3. Standard curve of P-ELISAs.

3.4. Specificity of P-ELISAs

The specificity of P-ELISAs, defined as CR, is shown in Table 2. Due to the similar structure of neonicotinoids, the competitive P-ELISA has 105.8%, 70.5%, 27.1%, 14.8% and 11.6% CRs to imidaclothiz, clothianidin, thiacloprid, nitenpyram and acetamiprid. Meanwhile, thanks to the two-site recognition pattern of anti-immunocomplex peptides, the noncompetitive P-ELISA greatly reduces the CRs of clothianidin, thiacloprid, nitenpyram and acetamiprid to <0.1%, 21.7%, and <0.1% and 6.8%. The CRs to imidaclothiz is a widespread problem in the immunoassays for imidacloprid [33]. Regrettably, the utilization of anti-immunocomplex peptides has not solved this problem either.

Table 2. CR of P-ELISAs with other neonicotinoids (n = 3).

Compound	Chemical Structure	Competitive P-ELISA		Noncompetitive P-ELISA	
		IC_{50} (ng/mL)	CR (%)	SC_{50} (ng/mL)	CR (%)
Imidacloprid	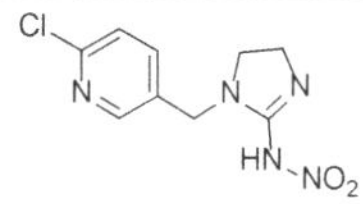	0.55	100.0	0.35	100.0
Imidaclothiz	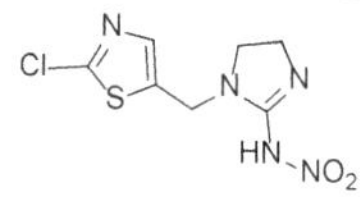	0.52	105.8	0.31	112.9
Thiacloprid	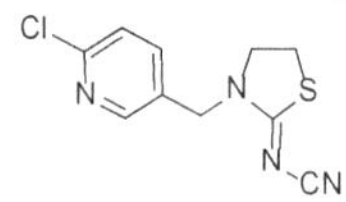	2.03	27.1	1.61	21.7
Acetamiprid	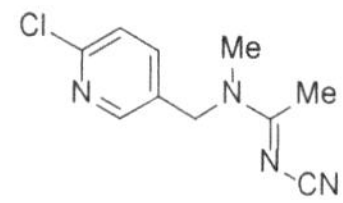	4.75	11.6	5.15	6.8
Clothianidin	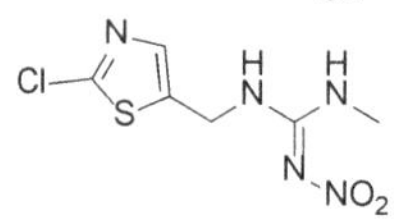	0.78	70.5	>10,000.00	<0.1
Nitenpyram	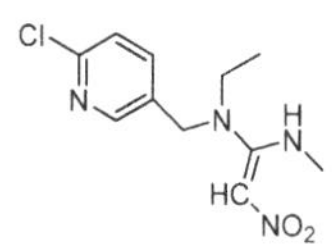	3.71	14.8	>10,000.00	<0.1
Thiamethoxam	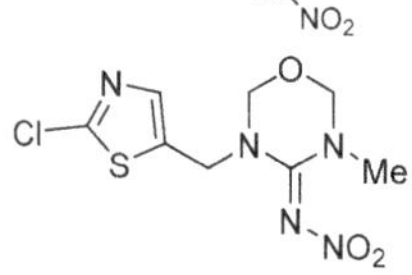	>10,000.00	<0.1	>10,000.00	<0.1
Dinotefuran	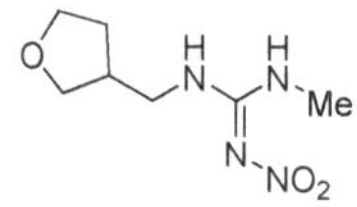	>10,000.00	<0.1	>10,000.00	<0.1

3.5. Recoveries of Spiked Samples

The matrix effect that would cause the false result has to be solved before testing actual samples. Owing to the high sensitivity of immunoassays, the matrix effects of extracts can simply be eliminated by the dilution method. To evaluate their influence, series diluted matrix extracts were used to replace the optimal buffer and run the dose-response curves of P-ELISAs. When the dilution fold produces a similar curve to the standard curve, the matrix effect is considered to be eliminated. As shown in Figures S2 and S3, the effect of matrix extracts on competitive P-ELISA can be eliminated after 2-fold dilution of paddy water, total 36-fold dilution of soil and pear, 72-fold dilution of orange and brown rice, and 144-fold dilution of cabbage, potato, and wheat, while the effect on noncompetitive P-ELISA can be eliminated after 2-fold dilution of paddy water, 36-fold dilution of potato and orange, 72-fold dilution of soil, wheat, cabbage, and pear, and 144-fold dilution of brown rice.

After the elimination of matrix effects, the IMI concentrations in the spiked samples were determined (Table 3). The average recoveries of competitive and noncompetitive P-ELISAs were 75.1–103.5% and 73.8–101.0%, with relative strand deviations (RSDs) of 1.7–7.8% and 2.0–9.3%, respectively. The results agreed with the IUPAC standard [34] regulation, "average recoveries among 70–120%, RSD ≤ 20%".

Table 3. Recoveries of spiked samples for IMI (n = 3).

Matrix	Spiked (ng/g)	Competitive P-ELISA			Noncompetitive P-ELISA		
		Measured ± SD (ng/g)	Average Recovery (%)	RSD (%)	Measured ± SD (ng/g)	Average Recovery (%)	RSD (%)
	2	1.81 ± 0.07	90.7	4.1	1.77 ± 0.07	88.5	4.1
Paddy water	4	3.69 ± 0.10	92.3	2.8	3.56 ± 0.23	88.9	6.4
	8	8.27 ± 0.49	103.4	5.9	7.78 ± 0.41	97.2	5.3
	50	42.81 ± 0.95	85.6	2.2	44.91 ± 2.51	89.8	5.6
Soil	100	90.02 ± 2.66	90.0	3.0	101.01 ± 4.01	101.0	4.0
	200	185.63 ± 4.11	92.8	2.2	186.59 ± 11.78	93.3	6.3
	50	38.29 ± 2.28	76.6	6.0	42.34 ± 3.41	84.7	8.1
Wheat	100	81.17 ± 1.42	81.2	1.7	84.52 ± 7.89	84.5	9.3
	200	181.40 ± 6.84	90.7	3.8	165.99 ± 11.78	83.0	7.1
	50	42.04 ± 2.40	84.1	5.7	45.39 ± 1.92	90.8	4.2
Brown rice	100	85.22 ± 6.64	85.2	7.8	89.86 ± 8.17	89.9	9.1
	200	173.29 ± 12.48	86.6	7.2	192.07 ± 6.66	96.0	3.5
	50	44.94 ± 1.89	89.9	4.2	40.47 ± 2.45	80.9	6.0
Cabbage	100	84.36 ± 3.23	84.4	3.8	81.74 ± 4.04	81.7	4.9
	200	164.56 ± 8.82	82.3	5.4	183.67 ± 5.59	91.8	3.0
	50	44.2 ± 3.04	88.4	6.9	46.79 ± 2.00	93.6	4.3
Potato	100	88.15 ± 4.15	88.2	4.7	86.62 ± 6.84	86.6	7.9
	200	182.38 ± 4.89	91.2	2.7	178.69 ± 8.36	89.3	4.7
	50	37.56 ± 1.38	75.1	3.7	38.28 ± 1.35	76.6	3.5
Pear	100	79.41 ± 2.85	79.4	3.6	85.37 ± 1.82	85.4	2.1
	200	173.86 ± 11.52	86.9	6.6	193.77 ± 3.94	96.9	2.0
	50	46.63 ± 2.11	93.3	4.5	36.91 ± 1.93	73.8	5.2
Orange	100	99.92 ± 6.06	99.9	6.1	85.36 ± 3.14	85.4	3.7
	200	207.07 ± 5.82	103.5	2.8	180.38 ± 10.15	90.2	5.6

3.6. Verification by HPLC

As a gold standard for pesticide residue detection, HPLC was used to further verify the accuracy of the proposed P-ELISAs. The blind samples of pear and soil were simultaneously detected by HPLC and the P-ELISAs. The results are shown in Figure 4. The correlation equations between competitive P-ELISA and HPLC are y = 1.11x + 0.65 (R^2 = 0.9739) for pear and y = 0.98x + 5.07 (R^2 = 0.9837) for soil, while y = 0.86x + 29.82 (R^2 = 0.9629) for pear and y = 0.94x + 8.13 (R^2 = 0.9964) for soil between noncompetitive P-ELISA and HPLC. The slope and R^2 of the equations are close to 1, which indicates a good correlation between HPLC and the immunoassays.

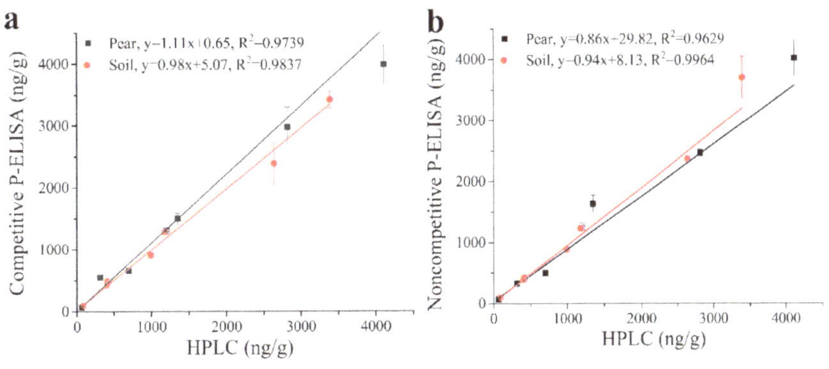

Figure 4. Blind samples tested by P-ELISAs and HPLC. (**a**) Competitive P-ELISA. (**b**) Noncompetitive P-ELISA.

4. Conclusions

In this study, thirty sequences of peptidomimetics and two sequences of anti-immunocomplex peptides for IMI were isolated from phage display peptide libraries, among which the anti-immunocomplex peptides were the first reported noncompetitive reagent for IMI.

The peptidomimetic 1-9-H and anti-immunocomplex peptide 2-1-H that showed the best sensitivities were utilized to develop competitive and noncompetitive P-ELISAs. The IC_{50} and LOD of competitive P-ELISA were 0.55 ng/mL and 0.3 ng/mL, compared with 0.35 ng/mL and 0.15 ng/mL for noncompetitive P-ELISA. The sensitivities of P-ELISAs were better than most reported immunoassays for IMI, which can be further improved by collaborating with other tracers. Moreover, the two-site recognition pattern of anti-immune complex peptides in noncompetitive P-ELISA greatly improved the specificity compared with competitive P-ELISA. As confirmed by the recoveries analysis and HPLC verification, the proposed P-ELISAs can detect IMI in agricultural and environmental samples accurately and reliably. IMI residue has always been the focus of attention due to IMIs wide use. So, we believe that the specific peptide sequences and methods reported here are helpful to detect IMI residue. Further, unlike chemical hapten, peptide ligands can be prepared using a biological expression, so it is attractive to explore the peptide ligands to derive novel patterns of immunoassay, such as label-free immunoassay.

Supplementary Materials: The following supporting information can be downloaded at: https://www.mdpi.com/article/10.3390/foods11203163/s1, Figure S1: Optimization of mAb concentration and phage input in noncompetitive P-ELISA; Figure S2: Matrix effects on competitive P-ELISA. Figure S3: Matrix effects of noncompetitive P-ELISA; Table S1: Phage titer among panning; Table S2: Optimization of mAb concentrations and phage inputs in competitive P-ELISA; Table S3: Comparison of reported ELISAs for IMI.

Author Contributions: Conceptualization, methodology, software, investigation, formal analysis, and writing original draft, T.Y.; investigation, methodology, software, and writing original draft, Y.D.; methodology, and data curation, Y.L.; methodology, Y.H.; validation, formal analysis, visualization, and supervision, M.W.; conceptualization, resources, writing review and editing, supervision, and project administration, X.H.; All authors have read and agreed to the published version of the manuscript.

Funding: This work was supported by the National Natural Science Foundation of China (Grant 31972311), Project funded by China Postdoctoral Science Foundation (Grant 2021M701745) and Jiangsu Planned Projects for Postdoctoral Research Funds (2021K103B).

Institutional Review Board Statement: Not applicable.

Informed Consent Statement: Not applicable.

Data Availability Statement: Data are contained within the article.

Conflicts of Interest: The authors declare no conflict of interest.

References

1. Yao, J.J.; Wang, Z.X.; Guo, L.L.; Xu, X.X.; Liu, L.Q.; Xu, L.G.; Song, S.S.; Xu, C.L.; Kuang, H. Advances in immunoassays for organophosphorus and pyrethroid pesticides. *Trends Analyt. Chem.* **2020**, *131*, 116022. [CrossRef]
2. Zhao, M.Q.; Wang, M.; Zhang, X.G.; Zhu, Y.A.; Cao, J.; She, Y.X.; Cao, Z.; Li, G.Y.; Wang, J.; Abd El-Aty, A.M. Recognition elements based on the molecular biological techniques for detecting pesticides in food: A review. *Crit. Rev. Food Sci. Nutr.* **2021**. [CrossRef] [PubMed]
3. Hua, X.D.; Eremin, A.S.; Liu, F.Q.; Wang, M.H. Antibody developments and immunoassays for organophosphorus chemicals: A review. *Curr. Org. Chem.* **2017**, *21*, 2640–2652. [CrossRef]
4. Jin, M.J.; Chen, G.; Du, P.F.; Zhang, C.; Cui, X.Y.; Gee, J.S.; She, Y.X.; Zheng, L.F.; Wang, S.S.; Shao, H.; et al. Developments on immunoassays for pyrethroid chemicals. *Curr. Org. Chem.* **2017**, *21*, 2653–2661. [CrossRef]
5. Vasylieva, N.; Ahn, K.C.; Barnych, B.; Gee, S.J.; Hammock, B.D. Development of an immunoassay for the detection of the phenylpyrazole insecticide fipronil. *Environ. Sci. Technol.* **2015**, *49*, 10038–10047. [CrossRef]
6. He, J.X.; Tao, X.W.; Wang, K.; Ding, G.C.; Li, J.; Li, Q.X.; Gee, S.J.; Hammock, B.D.; Xu, T. One-step immunoassay for the insecticide carbaryl using a chicken single-chain variable fragment (scFv) fused to alkaline phosphatase. *Anal. Biochem.* **2019**, *572*, 9–15. [CrossRef]
7. Wang, J.; Bever, C.R.S.; Majkova, Z.; Dechant, J.E.; Yang, J.; Gee, S.J.; Xu, T.; Hammock, B.D. Heterologous antigen selection of camelid heavy chain single domain antibodies against Tetrabromobisphenol A. *Anal. Chem.* **2014**, *86*, 8296–8302. [CrossRef]
8. Zhao, F.C.; Shi, R.R.; Liu, R.X.; Tian, Y.; Yang, Z.Y. Application of phage-display developed antibody and antigen substitutes in immunoassays for small molecule contaminants analysis: A mini-review. *Food Chem.* **2021**, *339*, 128084. [CrossRef]

9. Du, M.; Yang, Q.; Liu, W.M.; Ding, Y.; Chen, H.; Hua, X.D.; Wang, M.H. Development of immunoassays with high sensitivity for detecting imidacloprid in environment and agro-products using phage-borne peptides. *Sci. Total Environ.* **2020**, *723*, 137909. [CrossRef]
10. Rossotti, M.A.; Carlomagno, M.; González-Techera, A.; Hammock, B.D.; Last, J.; González-Sapienza, G. Phage anti-immunocomplex assay for clomazone: Two-site recognition increasing assay specificity and facilitating adaptation into an on-site format. *Anal. Chem.* **2010**, *82*, 8838–8843. [CrossRef]
11. Li, Y.S.; Zhang, G.Z.; Mao, X.; Yang, S.P.; De Ruyck, K.; Wu, Y.N. High sensitivity immunoassays for small molecule compounds detection-Novel noncompetitive immunoassay designs. *Trac-Trends Anal. Chem.* **2018**, *103*, 198–208. [CrossRef]
12. Shi, R.R.; Zhao, Z.L.; Wang, G.Q.; Zou, W.T.; Zhao, F.C.; Yang, Z.Y. Development of a noncompetitive magnetic-phage anti-immunocomplex assay for detecting of organophosphorus pesticides with a thiophosphate group. *Anal. Biochem.* **2022**, *646*, 114632. [CrossRef]
13. You, T.Y.; Ding, Y.; Chen, H.; Song, G.Y.; Huang, L.R.; Wang, M.H.; Hua, X.D. Development of competitive and noncompetitive immunoassays for clothianidin with high sensitivity and specificity using phage-displayed peptides. *J. Hazard. Mater.* **2022**, *425*, 128011. [CrossRef]
14. Lassabe, G.; Kramer, K.; Hammock, B.D.; González-Sapienza, G.; González-Techera, A. Noncompetitive homogeneous detection of small molecules using synthetic nanopeptamer-based luminescent oxygen channeling. *Anal. Chem.* **2018**, *90*, 6187–6192. [CrossRef]
15. Wang, L.; Liu, T.Z.; Liu, F.; Zhang, J.J.; Wu, Y.H.; Sun, H.W. Occurrence and profile characteristics of the pesticide imidacloprid, preservative parabens, and their metabolites in human urine from rural and urban china. *Environ. Sci. Technol.* **2015**, *49*, 14633–14640. [CrossRef]
16. Batikian, C.M.; Lu, A.; Watanabe, K.; Pitt, J.; Gersberg, R.M. Temporal pattern in levels of the neonicotinoid insecticide, imidacloprid, in an urban stream. *Chemosphere* **2019**, *223*, 83–90. [CrossRef]
17. Craddock, H.A.; Huang, D.; Turner, P.C.; Quirós-Alcalá, L.; Payne-Sturges, D.C. Trends in neonicotinoid pesticide residues in food and water in the United States, 1999–2015. *Environ. Health* **2019**, *18*, 7. [CrossRef]
18. Yang, W.; Carmichael, S.L.; Roberts, E.M.; Kegley, S.E.; Padula, A.M.; English, P.B.; Shaw, G.M. Residential agricultural pesticide exposures and risk of neural tube defects and orofacial clefts among offspring in the San Joaquin Valley of California. *Am. J. Epidemiol.* **2014**, *179*, 740–748. [CrossRef]
19. Carmichael, S.L.; Yang, W.; Roberts, E.; Kegley, S.E.; Padula, A.M.; English, P.B.; Lammer, E.J.; Shaw, G.M. Residential agricultural pesticide exposures and risk of selected congenital heart defects among offspring in the San Joaquin Valley of California. *Environ. Res.* **2014**, *135*, 133–138. [CrossRef]
20. Keil, A.P.; Daniels, J.L.; Hertz-Picciotto, I. Autism spectrum disorder, flea and tick medication, and adjustments for exposure misclassification: The CHARGE (CHildhood Autism Risks from Genetics and Environment) case–control study. *Environ. Health* **2014**, *13*, 3. [CrossRef]
21. Koureas, M.; Tsezou, A.; Tsakalof, A.; Orfanidou, T.; Hadjichristodoulou, C. Increased levels of oxidative DNA damage in pesticide sprayers in Thessaly Region (Greece). Implications of pesticide exposure. *Sci. Total Environ.* **2014**, *496*, 358–364. [CrossRef] [PubMed]
22. Hernández, A.F.; Casado, I.; Pena, G.; Gil, F.; Villanueva, E.; Pla, A. Low level of exposure to pesticides leads to lung dysfunction in occupationally exposed subjects. *Inhal. Toxicol.* **2008**, *20*, 839–849. [CrossRef] [PubMed]
23. Costa, C.; Silvari, V.; Melchini, A.; Catania, S.; Heffron, J.J.; Trovato, A.; De Pasquale, R. Genotoxicity of imidacloprid in relation to metabolic activation and composition of the commercial product. *Mutat. Res. Gen. Tox. En.* **2009**, *672*, 40–44. [CrossRef] [PubMed]
24. Li, K.; Li, Q.X. Development of an enzyme-linked immunosorbent assay for the insecticide imidacloprid. *J. Agric. Food Chem.* **2000**, *48*, 3378–3382. [CrossRef]
25. Lee, J.K.; Ahn, K.C.; Park, O.S.; Kang, S.Y.; Hammock, B.D. Development of an ELISA for the detection of the residues of the insecticide imidacloprid in agricultural and environmental samples. *J. Agric. Food Chem.* **2001**, *49*, 2159–2167. [CrossRef]
26. Kim, H.-J.; Shelver, W.L.; Li, Q.X. Monoclonal antibody-based enzyme-linked immunosorbent assay for the insecticide imidacloprid. *Anal. Chim. Acta.* **2004**, *509*, 111–118. [CrossRef]
27. Wang, R.M.; Wang, Z.H.; Yang, H.; Wang, Y.Z.; Deng, A.P. Highly sensitive and specific detection of neonicotinoid insecticide imidacloprid in environmental and food samples by a polyclonal antibody-based enzyme-linked immunosorbent assay. *J. Sci. Food Agric.* **2012**, *92*, 1253–1260. [CrossRef]
28. Hua, X.D.; Wang, L.M.; Li, G.; Fang, Q.K.; Wang, M.H.; Liu, F.Q. Multi-analyte enzyme-linked immunosorbent assay for organophosphorus pesticides and neonicotinoid insecticides using a bispecific monoclonal antibody. *Anal. Methods* **2013**, *5*, 1556–1563. [CrossRef]
29. Wang, Y.D.; Qin, J.a.; Zhang, J.; Jin, Z.Y.; Luo, J.Y.; Yang, M.H. Rapid screening of imidacloprid residue in grains and medicinal herbs: A newly designed hapten and monoclonal antibody. *J. Pharm. Biomed. Anal.* **2022**, *219*, 114931. [CrossRef]
30. Yang, J.C.; Yang, Q.; Deng, J.Q.; Tao, Z.X.; Hua, X.D.; Wang, M.H. Development of immunochromatographic assays for the detection of imidacloprid in soil chemical barrier. *Environ. Sci. Pollut. Res.* **2018**, *25*, 26617–26624. [CrossRef]
31. Reverberi, R.; Reverberi, L. Factors affecting the antigen-antibody reaction. *Blood Transfus.* **2007**, *5*, 227–240. [CrossRef]
32. Liu, Z.P.; Liu, J.F.; Wang, K.; Li, W.H.; Shelver, W.L.; Li, Q.X.; Li, J.; Xu, T. Selection of phage-displayed peptides for the detection of imidacloprid in water and soil. *Anal. Biochem.* **2015**, *485*, 28–33. [CrossRef]

33. Guo, Y.R.; Zou, R.B.; Si, F.F.; Liang, W.L.; Zhang, T.Y.; Chang, Y.Y.; Qiao, X.S.; Zhao, J.H. A sensitive immunoassay based on fluorescence resonance energy transfer from up-converting nanoparticles and graphene oxide for one-step detection of imidacloprid. *Food Chem.* **2021**, *335*, 127609. [CrossRef]
34. Krotxky, A.J.; Zeeh, B. Immunoassays for residue analysis of agrochemicals: Proposed guidelines for precision, standardization and quality control. *Pure Appl. Chem.* **1995**, *67*, 2065–2088. [CrossRef]

foods

MDPI

Article

An Alkyne-Mediated SERS Aptasensor for Anti-Interference Ochratoxin A Detection in Real Samples

Hao Wang [1,2], Lu Chen [1,2], Min Li [1,2], Yongxin She [3], Chao Zhu [1,2,*] and Mengmeng Yan [1,2,*]

1 Institute of Quality Standard and Testing Technology for Agro-Products, Shandong Academy of Agricultural Sciences, Jinan 250100, China
2 Shandong Provincial Key Laboratory Test Technology on Food Quality and Safety, Jinan 250100, China
3 Institute of Quality Standard and Testing Technology for Agro-Products, Chinese Academy of Agricultural Science, Beijing 100081, China
* Correspondence: ndytzhuchao@126.com (C.Z.); ynky202@163.com (M.Y.)

Abstract: Avoiding interference and realizing the precise detection of mycotoxins in complex food samples is still an urgent problem for surface-enhanced Raman spectroscopy (SERS) analysis technology. Herein, a highly sensitive and specific aptasensor was developed for the anti-interference detection of Ochratoxin A (OTA). In this aptasensor, 4-[(Trimethylsilyl) ethynyl] aniline was employed as an anti-interference Raman reporter to prove a sharp Raman peak (1998 cm^{-1}) in silent region, which could avoid the interference of food bio-molecules in 600–1800 cm^{-1}. 4-TEAE and OTA-aptamer were assembled on Au NPs to serve as anti-interference SERS probes. Meanwhile, Fe_3O_4 NPs, linked with complementary aptamer (cApts), were applied as capture probes. The specific binding of OTA to aptamer hindered the complementary binding of aptamer and cApt, which inhibited the binding of SERS probes and capture probes. Hence, the Raman responses at 1998 cm^{-1} were negatively correlated with the OTA level. Under the optimum condition, the aptasensor presented a linear response for OTA detection in the range of 0.1–40 nM, with low detection limits of 30 pM. In addition, the aptasensor was successfully applied to quantify OTA in soybean, grape and milk samples. Accordingly, this anti-interference aptasensor could perform specific, sensitive and precise detection of OTA in real samples, and proved a reliable reference strategy for other small-molecules detection in food samples.

Keywords: SERS; 4-TEAE; aptamer; silent region; OTA; anti-interference

Citation: Wang, H.; Chen, L.; Li, M.; She, Y.; Zhu, C.; Yan, M. An Alkyne-Mediated SERS Aptasensor for Anti-Interference Ochratoxin A Detection in Real Samples. *Foods* **2022**, *11*, 3407. https://doi.org/10.3390/foods11213407

Academic Editors: Marco Beccaria

Received: 14 September 2022
Accepted: 25 October 2022
Published: 28 October 2022

Publisher's Note: MDPI stays neutral with regard to jurisdictional claims in published maps and institutional affiliations.

Copyright: © 2022 by the authors. Licensee MDPI, Basel, Switzerland. This article is an open access article distributed under the terms and conditions of the Creative Commons Attribution (CC BY) license (https://creativecommons.org/licenses/by/4.0/).

1. Introduction

Ochratoxin A (OTA) is a naturally occurring mycotoxin that is one of the most toxic contaminants in dairy products, crops and alcoholic beverages, with a long half-life and high chemical stability [1]. OTA has nephrotoxicity, hepatotoxicity, and immunotoxicity for humans; hence, many countries have regulations for OTA limits. According to European Commission regulations, the maximum level for OTA in grains, grape juice and wine are 5 μg/kg, 2 μg/kg and 2 μg/kg, respectively [2]. To avoid the hazards of OTA for human health, quantifying the contaminated level of OTA in agricultural products is of great significance. In recent decades, instrumental methods [3] and antibody-based methods [4] have developed to be prevalent methods for OTA detection. However, expensive instruments and the demand for trained technicians limit the development of instrumental methods. In addition, high prices and the stringent conditions of transportation and storage limit the practical applicability of antibody-based methods [5]. Therefore, it is highly desirable to explore simple, low-cost and sensitive OTA detection methods.

In recent years, surface-enhanced Raman spectroscopy (SERS)-based biosensor has acted as a promising alternative in food safety, due to its high sensitivity, specificity and multiplexing detection ability [3,6]. As the core component of SERS-based biosensor, SERS probes are normally composed of metal nanostructures (enhanced substrates), such as gold nanoparticles (Au NPs) and Raman reporters (RRs) [7]. RRs (such as 4-mercaptobenzoic acid, 4-mercaptopyridine, 4-nitrothiophenol, etc.) are easily and firmly attached to the surface of metal nanostructures, which could enhance the inherent Raman emissions of RRs to 10^6–10^{14} times under the action of electromagnetic mechanism and chemical enhancement [8]. SERS probes provide clear, strong and steady Raman signals of RRs, which could perform high sensitivity and high fidelity analysis [9]. In recent years, some biosensors based on SERS probes for the quantitative analysis of mycotoxins (such as zearalenone and aflatoxin B1) [10,11] have been developed. However, there are still some problems that hindered the practical application of SERS-based biosensors in food analysis. In particular, the Raman emission of food biomolecules and traditional RRs are formed in the region of <1800 cm^{-1}; resulting Raman interference could seriously affect the accuracy of the detection results. Therefore, it is urgent to develop an anti-interference RRs for food analysis [12,13].

Recently, researchers have reported some new RRs, which contain C≡C, C=O and C≡N groups [14–16]. Those RRs displayed narrow peak in the silent region (1800–2800 cm^{-1}), which can effectively avoid the Raman interference of food biomolecules. As new RRs are created, those anti-interference RRs have been applied in SERS technology, such as biosensors, cell imaging, biomedicine and other fields [17–20]. As a major classification of anti-interference RRs, alkynyl-containing RRs have unique Raman shift in the silent region. The exploration of alkynyl-containing RRs is helpful to the development of Raman silent region tags and expand the application potential of anti-interference RRs. However, thus far, there are no reports about applying alkynyl compound as anti-interference RRs in the food safety field.

Therefore, we present an alkyne-mediated SERS-based aptasensor for OTA detection in real samples (soybean, grape and milk). As shown in Figure 1, 4-[(Trimethylsilyl) ethynyl] aniline (4-TEAE) was applied as anti-interference Raman reporter; gold nanoparticles (Au NPs) linked with aptamers and 4-TEAE were used as Raman probes (4-TEAE/AuNPs/Apt). Meanwhile, Fe_3O_4 nanoparticles (Fe_3O_4 NPs) conjugated with cApt served as capture probes (Fe_3O_4 NPs/cApt). OTA could specifically interact with aptamer, which inhibited the conjunction between Raman probes and capture probes. Thus, the Raman intensity of 4-TEAE/AuNPs was inverse, correlating with the concentration of OTA. In order to verify the practicability of the aptasensor, we detected OTA in soybean, grape and milk samples, and compared the recovery rates with HPLC-MS/MS.

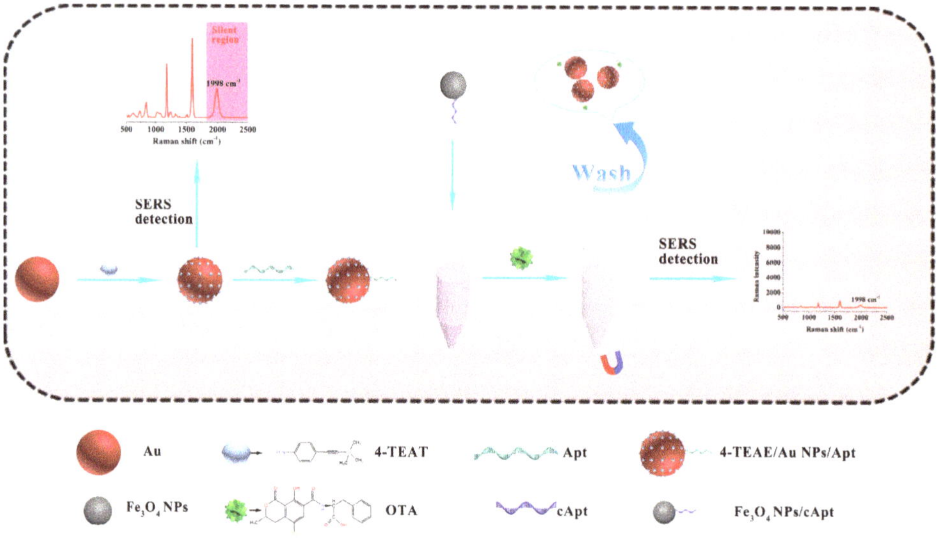

Figure 1. Schematic representation of the principle for the alkyne-mediated SERS aptasensor for anti-interference Ochratoxin A detection.

2. Materials and Methods

2.1. Materials and Reagent

Ochratoxin A, chloroauric Acid ($HAuCl_4$), trisodium citrate, 4-[(Trimethylsilyl) ethynyl] aniline, Tris (2-carboxyethy1) phosphine (TCEP), phosphate buffer saline (PBS), ethyl-3-(3-dimethylaminopropyl) carbodiimide hydrochloride (EDC), and N-Hydroxysuccinimide (NHS) were purchased from Aladdin Industrial Inc (Shanghai, China). Fe_3O_4 nanoparticles bonded with -COOH (10 mg/mL) were bought from Sigma-Aldrich (Beijing, China).

According to previous research [21], the sequences of aptamer and cApt are as follows: 5′-SH-GATCGGGTGTGGGTGGCGTAAAGGGAGCATCGGACA-3′ (aptamer), 5′-NH2-$(CH_2)_6$-CCTTTACGCCACCCACACCCGATC-3′ (cApt). All synthetic nucleotide sequences were acquired from Sangon Biotech Co., Ltd. (Shanghai, China)

2.2. Instruments

The synthesis of Au NPs was carried out by Magnetic electric heating sleeve (IKA, staufen, Germany). The recording and comparison of Raman spectrums were performed by miniature handheld Raman spectrometer (QEPro, Ocean Insight, Dunedin, FL, USA), and the laser power and laser wavelength of spectrometer were 390-410 mW and 785 nm, respectively. UV-vis spectra (synergy HTX, Biotek, Winooski, VT, USA) was applied to get ultraviolet visible (UV) absorbance. The morphology of nanoparticles were characterized by Transmission electron microscope (TEM) (JEM1200EX, JEOL, TKY, Nagoya, Japan). The chemical composition and relative content of the probes were obtained by the energy-dispersive X-ray (EDX) spectroscopy facility of scanning electron microscopy (SEM) (SU8020, Hitachi, TKY, Japan). Ultra-pure water was prepared by purification system (Millipore, Bedford, MA, USA) and used throughout this work.

2.3. Preparation of SERS Probe

Au NPs (25 nm) were synthesized by the sodium citrate reduction method [22]. First, 100 mL ultra-pure water was heated to boiling with a heating sleeve; 1 mL (0.01 g/mL) trisodium citrate solution was then added. After 1 min, 0.1 mL (0.1 g/mL) chloroauric acid solution was added to the boiling solution, and the color of the boiling liquid quickly changed from colorless to transparent wine red. After the color of Au NPs solution was stable and no longer changed, heating ceased and continuous stirring was performed for 20 min. After naturally cooling to room temperature, the Au NPs solution was stored at 4 °C for further use.

OTA aptamer was dissolved to 100 μM with TCEP solution and then diluted to 10 μM with PBS solution for further use. The 0.87 mL Au NPs solution was mixed with 100 μL 4-TEAE in a scroll oscillator for 1 h. The resulted solution was centrifuged under 6000 rpm for 15 min and washed with ultra-pure water three times to obtain 4-TEAE fixed Au NPs (Au NPs-4-TEAE). Then, 30 μL (10 μM) OTA aptamer was co-incubated with Au NPs-4-TEAE solution at room temperature for 4 h. Finally, the Au NPs-4-TEAE/Apt solution was centrifuged under 6000 rpm for 15 min, washed with ultra-pure water three times and re-dissolved in PBS solution to obtain anti-biomolecule interference SERS probes, and stored at 4 °C.

2.4. Preparation of Capture Probes

The surface of MNPs was immobilized with cApt and the capture probes were prepared. Briefly, 100 μL MNPs (10 mg/mL) solution was mixed with 7.9 mL ultra-pure water; 1 mL EDC (16 mM) and 1 mL NHS (4 mM) were added to protect the carboxyl groups on the surface of MNPs. After 30 min, MNPs solution was separated by magnet, washed with ultra-pure water three times, and then redissolved to 9.7 mL with PBS solution. Subsequently, 0.3 mL of cApt (10 μM) solution was added to the carboxyl-protected MNPs solution and vibrated for 1 h to prepare capture probes. After magnetic separation and removing the excess cApt, the capture probes were re-dissolved by PBS solution and stored at 4 °C for subsequent use.

2.5. Interference-Free Aptasensors for OTA Detection

Firstly, 1 mg/mL OTA acetonitrile solution was diluted with ultra-pure to 0.1–70 nM for further use. Secondly, a competitive binding system was constructed to link the OTA concentration with the SERS intensity at 1998 cm^{-1}. Briefly, 500 μL anti-interference SERS probe and 500 μL capture probe were mixed for 15 min, and then 20 μL OTA solution of different concentrations were added to the mixed solution and incubated with vibration for 25 min to fully react. Finally, the SERS probes that were bound to the capture probes were removed by magnetic separation and washed with ultra-pure water two times, and the remaining probes were suspended in the 1 mL of ultra-pure water for SERS measurement.

Finally, the SERS response at different OTA concentrations was measured by miniature handheld Raman spectrometer (laser wavelength = 785 nm and spectrometer power = 390–410 mW), and the response results to different OTA concentrations were obtained by the statistical results of three parallel measurements.

2.6. Procedures for OTA Residues Detection in Real Samples

To verify the practicability of the aptasensor, soybean, grape and milk were purchased from the supermarket and tested as real food samples. The pretreatment methods of food samples were referred to in previous reports [23–25]. The actual samples were added 1 nM, 5 nM and 10 nM of OTA, respectively. The extracted samples were added to the aptasensor system to complete the detection.

2.7. Specificity Analysis

In this work, AFB1, aflatoxin M1 (AFM1), zearalenone (ZEN), deoxynivalenol (DON) and trichosporene 2 (TMAE 2) were selected to verify the specificity of the aptasensor. Under the same conditions, these mycotoxins of 300 nM were simultaneously detected by the aptasensor with the OTA of 30 nM. The assays were repeated three times, and the SERS intensities at 1998 cm^{-1} were compared.

2.8. Statistical Analysis

All results were obtained from cubic parallel experiments, and standard deviation (SD) were represented with error bars one. Statistical analysis was performed with Origin 9.5. Ultimately, the limit of detection (LOD) was computed as LOD = 3 × SD/blank.

3. Results and Discussion

3.1. Synthesis and Characterization of 4-TEAE/Au NPs

In this study, 4-TEAE was applied as anti-interference Raman reporter to prove a strong and stable Raman signal at 1998 cm^{-1}, with the following reasoning: (1) Alkynyl linking with benzene ring presents sharp Raman scattering as the large Raman cross section of aromatic ring [26]. (2) 4-TEAE possesses a distinct Raman scattering at 1998 cm^{-1} due to the stretch of C≡C bond, achieving the anti-interference OTA detection in the complex samples. The Raman shifts of other Raman peaks of 4-TEAE, such as 1601 cm^{-1} and 1174 cm^{-1}, were not adopted because of the interference of the food biomolecules in the range of <1800 cm^{-1} (Figure 2d).

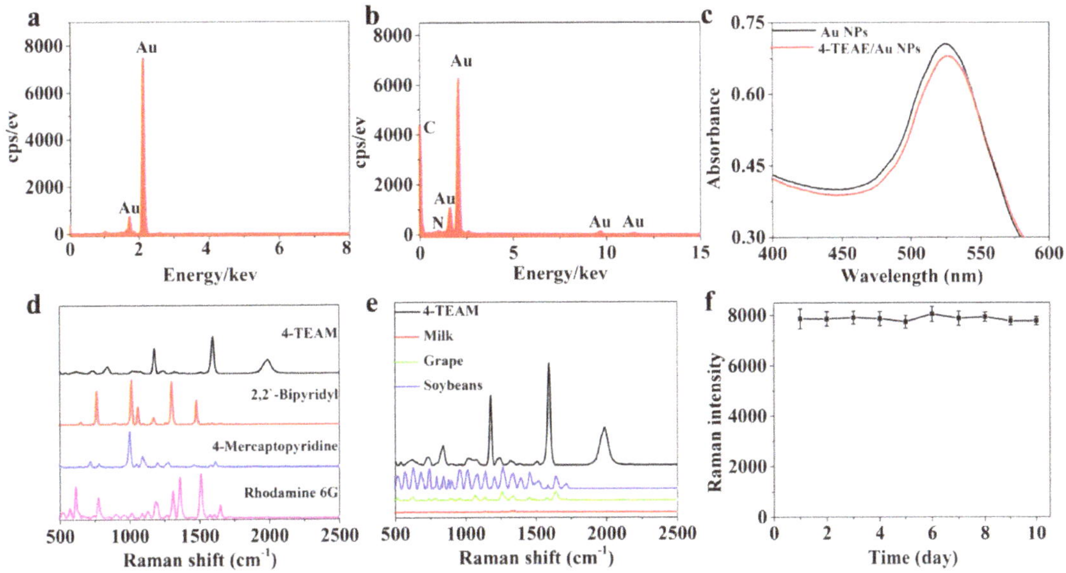

Figure 2. Characterization of the anti-interference SERS tag: (**a**) EDS spectroscopy of Au NPs. (**b**) EDS spectroscopy of 4-TAAE/Au NPs. (**c**) UV−vis spectrum of 4-TEAE/Au NPs (red), Au NPs (black). (**d**) Raman scatterings of conventional RRs and anti-interference SERS probe (black). (**e**) Raman scatterings of food samples and anti-interference SERS tag (black). (**f**) Stability of SERS tag for ten days.

The sharp Ramam scattering in 1998 cm^{-1} contributed to the combination of SERS enhanced substrate (Au NPs) and Raman reporter (4-TEAE). TEM was applied to image the combination of 4-TEAE/Au NPs. The synthesized Au NPs were spherical, with a particle size of 25 nm (Figure S1a). After linked with 4-TEAE, the nanoparticles were wrapped by a shadow with the thickness of 1.5 nm (Figure S1b). The combination of Au NPs and 4-TEAE was further identified by UV–vis spectra and EDX. UV–vis spectra confirmed that the absorption peak of Au NPs was red shifted after being coated by 4-TEAE. As shown in Figure 2c, the absorption peak of Au NPs presented at 524 nm, while the absorption peak of 4-TEAE/Au NPs was located at 526 nm. As revealed by Figure 2a,b, the EDX spectroscopy of Au NPs only possessed the peaks of Au, while the Au, C and N elements were co-existent in the 4-TEAE/Au NPs sample.

In order to verify the anti-interference ability of 4-TEAE, three traditional RRs (2,2′-Bipyridyl, Rhodamine 6G, 4-mercaptopyridine) were selected as Interfered control. The Raman scatterings of these RRs and food biomolecules were overlapping in the range of <1800 cm^{-1}. While 4-TEAE presented a distinct Raman signal in the silent region. 2,2′-Bipyridyl was chosen as the representative of traditional RRs and tested for the anti-interference ability of 4-TEAE. As shown in Figure 2d,e, the characteristic peaks of 2,2′-Bipyridyl (766 cm^{-1}, 1015 cm^{-1}, 1306 cm^{-1} and 1428 cm^{-1}) were severely overlapped with the spectral bands of Rhodamine 6G and 4-mercaptopyridine. Meanwhile, traditional RRs were interfered by food samples, especially soybeans and grapes. By contrast, 4-TEAE possessed a highly distinguished and stable peak at 1998 cm^{-1}, which gifted 4-TEAE superior anti-interference ability to the traditional RRs.

In summary, anti-interference SERS tags were successfully synthesized. To test the stability of anti-interference SERS tags, three portions of freshly prepared 4-TEAE/Au NPs solution were detected by Raman spectrometer at the same time every day for 10 days. As shown in Figure 2f, the anti-interference SERS tags presented a slight fluctuation in the characteristic peak (1998 cm^{-1}), with a Raman intensity deviation (RID) of less than 6%. Thus, the proposed anti-interference SERS tags were stable enough for OTA detection.

3.2. Characterization of SERS Probe and Capture Probe

3.2.1. Characterization of SERS Probe

SERS tags link with special single-stranded DNA (OTA-aptamer), which can specifically interact with OTA, gifting SERS probes with specific recognition ability to OTA. To achieve this, 4-TEAE/Au NPs was coated with aptamer via Au-S bond. As shown in Figure 3b, the EDX spectroscopy of 4-TEAE/Au NPs presented multiple peaks, which represented Au, C, N and S elements, respectively. Therefore, the SERS probes with specific recognition of OTA were successfully produced. The result of UV-vis spectra also confirmed the preparation of SERS probes. As shown in Figure 3a, the maximum absorption of Au was red shift from 526 nm to 530 nm, while 4-TEAE/Au NPs were linked with OTA-aptamers.

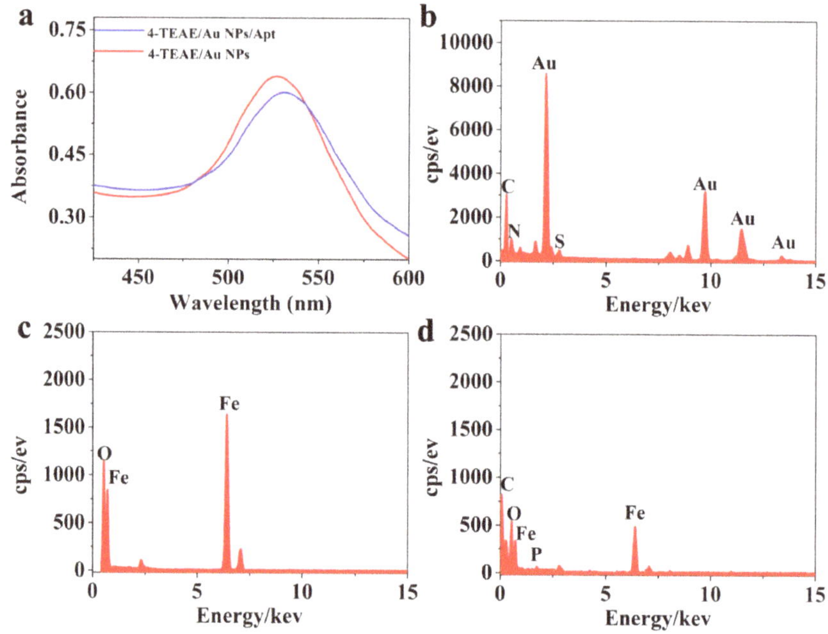

Figure 3. (**a**) UV−vis spectrum of 4-TEAE/Au NPs (red) and 4-TEAE/Au NPs/Apt (blue). EDS spectroscopy of (**b**) 4-TEAE/Au NPs/Apt, (**c**) Fe_3O_4 NPs and (**d**) and Fe_3O_4 NPs/cApt.

3.2.2. Characterization of Capture Probes

Fe_3O_4 NPs were selected as adsorption substrate, due to the large surface area, and it was easily separated and aggregated by external magnet. Furthermore, highly stable and biocompatible Fe_3O_4 NPs combined with cApt to form efficient and sensitive capture probes. TEM was applied to describe the form and size of Fe_3O_4 NPs. As shown in Figure S2, the Fe_3O_4 NPs are spherical, with an average size of 20 nm. Meanwhile, EDX spectroscopy provided sufficient proof for cApt binding to Fe_3O_4 NPs. As shown in Figure 3c,d, the characteristic peaks of Fe and O appeared in the spectrum of Fe_3O_4 NPs sample, while the characteristic peaks of Fe, O, C and P were observed in the spectra of Fe_3O_4 NPs/cApt. Thus, the capture probe with strong separation ability and high specificity was successfully fabricated.

3.3. Optimization of Experiment Conditions

The conditions of probes synthesis and competitive reaction conditions are vital to the sensitivity and specificity of aptasensor. Five key factors of aptasensor (concentration of 4-TEAE, concentration of aptamer, concentration of cApt, pH, and time of competitive reaction) were optimized to the best performance of OTA detection.

3.3.1. Optimization of Probe Synthesis

Firstly, the concentrations of 4-TEAE were altered from 10 to 80 μM, and the Raman intensities of different 4-TEAE concentrations at 1998 cm^{-1} were obtained by counting the results of three experiments. As shown in Figure 4a, the Raman intensity of 4-TEAE at 1998 cm^{-1} increased to the maximum when the concentration of 4-TEAE increased from 10 μM to 50 μM, while it reduced as the concentration of 4-TEAE increased to 80 μM. It could be clearly distinguished in Figure 4a, when the concentration of 4-TEAE reached 50 nM, the Raman intensity at 1998 cm^{-1} was much higher than that of other concentrations. Thus, the optimum concentration of 4-TEAE was selected as 50 nM.

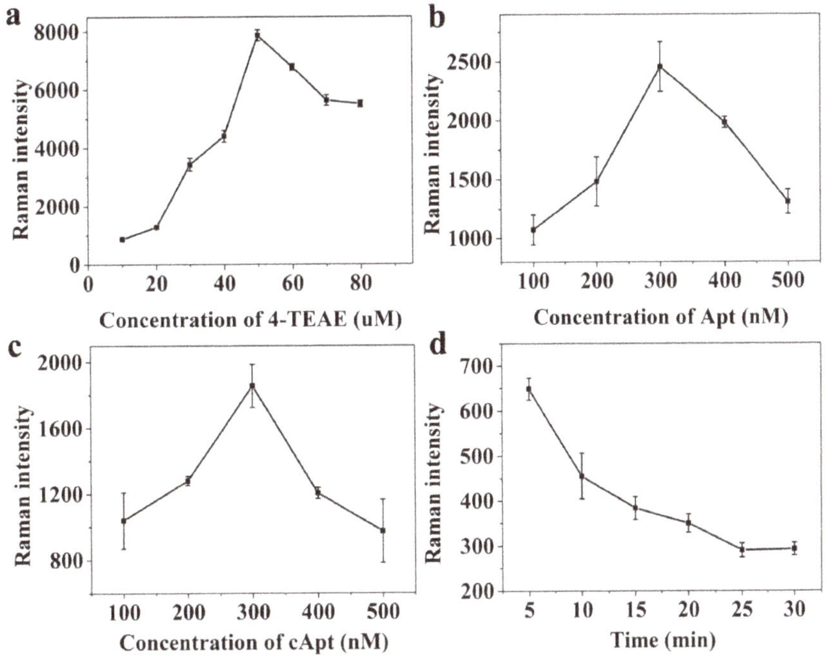

Figure 4. Raman intensity detected at 1998 cm^{-1} for the optimization of (**a**) 4-TEAE concentration, (**b**) Apt concentration, (**c**) cApt concentration and (**d**) competitive reaction time.

To investigate the effect of aptamer concentration on aptsensor performance, SERS probes with different concentrations (100–500 nM) of aptamers were combined with capture probes and separated by external magnets. Three parallel SERS intensities among different aptamer concentrations (100–500 nM) were counted and compared. In Figure 4b, the Raman intensity at 1998 cm^{-1} increased when the concentration of aptamer increased in the range of 100–300 nM, while it declined as the concentration of aptamer increased from 300 to 500 nM. When the level of aptamer was too little or reached excessive levels, the aptasensor performances were much lower than that at 300 nM. Therefore, the optimum aptamer concentration was 300 nM.

Finally, the optimal load of cApt to capture probes was analyzed by comparing the SERS performances of capture probes with different cApt concentrations. As depicted in

Figure 4c, the Raman intensity continuously enhanced until cApt concentration increased to 300 nM, while it decreased as the cApt concentration increased to 500 nM. Hence, the optimal concentration of cApt was set as 300 nM.

3.3.2. Optimization of Competitive Reaction Conditions

The time of competitive reaction has an important effect on aptasensor performance. In order to get the optimal reaction time, a series of reaction times (5, 10, 15, 20, 25, 30 min) were selected to determine the best reaction time. As shown in Figure 4d, with the increase of time, more SERS probes were separated from capture probes due to the presence of OTA. The Raman intensity of SERS probe showed a downward trend, and the Raman intensity was stable at 25 min, which was recorded as the optimal reaction time.

pH and temperature are the other two key factors of the competitive reaction, which, in turn, become the key factors affecting the sensitivity of the aptasensor. As shown in Figure S3, with the increase of OH^-, more bindings of SERS probes and capture probes were destroyed by OTA. Raman intensity declined, and reached the minimum at the concentration of H^+ was 10^{-7} M. In addition, owing to the high temperature inflicting damage to probes, referring to the relevant experiments [27,28], we determined that the competitive reaction should be carried out at 37 °C. Hence, a pH value of 7 and a temperature of 37 °C were the optimal conditions for the competitive reaction.

3.4. *Analytical Performance of the Aptasensor*

The aptasensor was performed for OTA detection under the optimal conditions, and the Raman responses to different concentrations of OTA were measured three times. Figure 5a depicted the negative correlation between Raman intensity and OTA concentration. In Figure 5c, the Raman response of 4-TEAE decreased with the concentration of OTA increasing from 0 to 40 nM. Figure 5b showed that the Raman intensity at 1998 cm^{-1} possessed a good linear relationship with the OTA concentration in the range of 0.1–40 nM. The correction curve is $y = 2152.14 - 46.73x$, $R^2 = 0.991$, with the LOD of 30 pM.

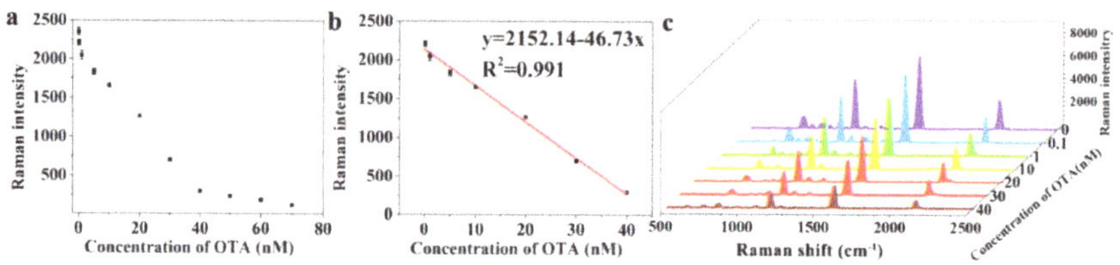

Figure 5. Analytical performance of the aptasensor: (**a**) SERS intensity difference at the peak of 1998 cm^{-1} with the different OTA concentrations, (**b**) linear relationship between SERS intensity and OTA concentration, (**c**) SERS spectra for OTA detection in the concentration range of 0–40 nM.

3.5. *Quantification of OTA in Real Samples*

To validate the practicability of the aptasensor, OTA detection was performed in real samples (soybean, grape and milk) for the recovery rate. In Table 1, different concentrations of OTA were added to those samples; the samples were pre-treated according to Section 2.6 and a recovery assay was carried out. The recovery rates of the OTA in these samples were found to be 83.3–100.8%, with relative standard deviations (RSDs) of 1.5–8.3%. The obtained results were in good agreement with those determined by the HPLC-MS/MS method, indicating that the aptasensor presents great practicability and reliability for OTA detection in real samples.

Table 1. Results of OTA Residues Detection in Real Samples by Aptasensor.

		SERS		HPLC−MS/MS	
Sample	**Added (nM)**	**Found (nM)**	**Recoveries/RSD (%)**	**Found (nM)**	**Recoveries/RSD [1] (%)**
soybean	1	0.951	95.1/2.1	0.991	99.1/8.1
	5	4.68	93.6/3.5	4.97	99.4/5.2
	10	10.08	100.8/5.3	9.87	98.7/4.1
grape	1	0.997	99.7/2.8	0.981	98.1/5.2
	5	4.56	91.2/1.5	4.86	97.2/5.4
	10	0.954	95.4/4.1	10.2	100.2/4.3
milk	1	0.938	93.8/2.7	1.03	103.0/4.5
	5	4.165	83.3/8.3	5.25	105.4/7.3
	10	9.42	94.2/3.1	10.1	100.1/3.9

[1] RSDs: relative standard deviation.

3.6. Selectivity

In this work, the specificity of the aptasensor was verified by comparing the Raman responds of 300 nM of mycotoxins (AFB1, AFM1, ZEN, DON and TMAE2) and OTA (30 nM) under the same conditions. As shown in Figure 6, the Raman responds of the mycotoxins at 1998 cm^{-1} were significantly higher than that of OTA, indicating the aptasensor possessed a marked affinity to OTA.

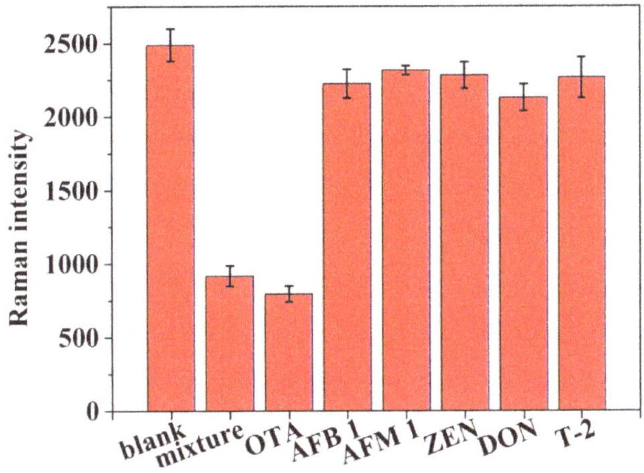

Figure 6. Selectivity evaluation of the ant-interference aptasensor for OTA detection (30 nM) against other mycotoxins (300 nM).

The performance of our aptasensor were also compared to those of previously proposed analysis methods [29–34]. Table S1 revealed that the proposed aptasensor for OTA detection was a novel, practical, and accurate method.

4. Conclusions

In this study, a novel anti-interference aptasensor was developed for OTA detection using 4-TEAE/Au NPs/Apt as SERS probes, and Fe_3O_4 NPs/cApt as capture probes. The RRs (4-TEAE) possessed a strong and stable Raman scattering at 1998 cm^{-1}, effectively avoiding the interference of the biomolecules in food. The specific interaction of OTA and aptamer provided high specificity and sensitivity for aptamer biosensor. The response of the sensor to OTA can be obtained through a miniature handheld Raman spectrometer, which has flexible application and avoids the limitations of complex operating conditions. Under the optimal conditions, the developed aptasensor presented a linear detection range of 0.1–40 nM, with a LOD of 30 pM to OTA. Aside from this, the results of aptasensor for

OTA detection in soybean, grape and milk was also tested, and satisfactory recoveries were obtained. In summary, the anti-interference aptasensor showed great performance in OTA detection and could be applied to enact precise detection of OTA residues in food samples.

Supplementary Materials: The following supporting information can be downloaded at: https://www.mdpi.com/article/10.3390/foods11213407/s1. Additional experimental results: TEM images and of Au NPs, 4-TEAE/Au NPs and Fe_3O_4 NPs; optimization of pH. The comparision of our aptasensor with others.

Author Contributions: H.W.: investigation; writing—original draft; formal analysis; data curation. L.C.: investigation. M.L.: investigation. Y.S.: investigation. C.Z.: methodology; formal analysis. M.Y.: funding acquisition; supervision; conceptualization; project administration; writing—review and editing. All authors have read and agreed to the published version of the manuscript.

Funding: This work was supported by the National Natural Science Foundation of China (32102088, 32172309), the Natural Science Foundation of Shandong Province (Grant No. ZR2022QB207), Supported by the Taishan Scholars Program, the Young Elite Scientist Sponsorship Program of Beijing Association for Science and Technology, the Agricultural Scientific and Technological Innovation Project of Shandong Academy of Agricultural Sciences (CXGC2018E05, CXGC2021B14, and CXGC2022E05), the Focus on Research and Development Plan in Shandong Province (No. 2020CXGC010804).

Data Availability Statement: Data are contained within the article.

Conflicts of Interest: The authors declare that they have no known competing financial interests or personal relationships that could have appeared to influence the work reported in this paper.

References

1. Malir, F.; Ostry, V.; Pfohl-Leszkowicz, A.; Malir, J.; Toman, J. Ochratoxin A 50 Years of Research. *Toxins* **2016**, *8*, 191. [CrossRef] [PubMed]
2. Smith, M.C.; Madec, S.; Coton, E.; Hymery, N. Natural Co-Occurrence of Mycotoxins in Foods and Feeds and Their in vitro Combined Toxicological Effects. *Toxins* **2016**, *8*, 94. [CrossRef] [PubMed]
3. Li, P.; Zhang, Z.; Hu, X.; Zhang, Q. Advanced hyphenated chromatographic-mass spectrometry in mycotoxin determination: Current status and prospects. *Mass Spectrom. Rev.* **2013**, *32*, 420–452. [CrossRef] [PubMed]
4. Liu, B.H.; Hsu, Y.-T.; Lu, C.C.; Yu, F.-Y. Detecting aflatoxin B1 in foods and feeds by using sensitive rapid enzyme-linked immunosorbent assay and gold nanoparticle immunochromatographic strip. *Food Control* **2013**, *30*, 184–189. [CrossRef]
5. Zhao, X.; Shen, H.; Huo, B.; Wang, Y.; Gao, Z. A novel bionic magnetic SERS aptasensor for the ultrasensitive detection of Deoxynivalenol based on "dual antennae" nano-silver. *Biosens. Bioelectron.* **2022**, *211*, 114383. [CrossRef]
6. Laing, S.; Gracie, K.; Faulds, K. Multiplex in vitro detection using SERS. *Chem. Soc. Rev.* **2016**, *45*, 1901–1918. [CrossRef]
7. Liu, Y.Q.; Zhu, W.; Hu, J.M.; Shen, A.G. Recent advances in plasmonic Prussian blue-based SERS nanotags for biological application. *Nanoscale Adv.* **2021**, *3*, 6568–6579. [CrossRef]
8. Jiang, Y.; Sun, D.W.; Pu, H.; Wei, Q. Surface enhanced Raman spectroscopy (SERS): A novel reliable technique for rapid detection of common harmful chemical residues. *Trends Food Sci. Technol.* **2018**, *75*, 10–22. [CrossRef]
9. Ganbold, E.O.; Lee, C.M.; Cho, E.M.; Son, S.J.; Kim, S.; Joo, S.W.; Yang, S.I. Subnanomolar detection of ochratoxin A using aptamer-attached silver nanoparticles and surface-enhanced Raman scattering. *Anal. Methods* **2014**, *6*, 3573–3577. [CrossRef]
10. Liu, J.; Hu, Y.; Zhu, G.; Zhou, X.; Jia, L.; Zhang, T. Highly sensitive detection of zearalenone in feed samples using competitive surface-enhanced Raman scattering immunoassay. *J. Agric. Food Chem.* **2014**, *62*, 8325–8332. [CrossRef]
11. Fang, C.; Wei, C.; Xu, M.; Yuan, Y.; Gu, R.; Yao, J. Ni@ Au nanoparticles for surface enhanced Raman spectroscopy based ultrasensitive magnetic immunoassay on aflatoxin B 1. *RSC Adv.* **2016**, *6*, 61325–61333. [CrossRef]
12. Wang, Y.; Yan, B.; Chen, L. SERS tags: Novel optical nanoprobes for bioanalysis. *Chem. Rev.* **2013**, *113*, 1391–1428. [CrossRef] [PubMed]
13. Sun, F.; Ella-Menye, J.R.; Galvan, D.D.; Bai, T.; Hung, H.C.; Chou, Y.N.; Zhang, P.; Jiang, S.; Yu, Q. Stealth surface modification of surface-enhanced Raman scattering substrates for sensitive and accurate detection in protein solutions. *ACS Nano* **2015**, *9*, 2668–2676. [CrossRef] [PubMed]
14. Hu, F.; Zeng, C.; Long, R.; Miao, Y.; Wei, L.; Xu, Q.; Min, W. Supermultiplexed optical imaging and barcoding with engineered polyynes. *Nat. Methods* **2018**, *15*, 194–200. [CrossRef] [PubMed]
15. Wei, L.; Yu, Y.; Shen, Y.; Wang, M.C.; Min, W. Vibrational imaging of newly synthesized proteins in live cells by stimulated Raman scattering microscopy. *Proc. Natl. Acad. Sci. USA* **2013**, *110*, 11226–11231. [CrossRef]
16. Wei, X.; Sun, Y.; Liu, C.; Li, Z.; Zou, X.; Zhang, D.; Zhang, W.; Shi, J.; Huang, X.; Li, Y. A nitrile-mediated SERS aptasensor coupled with magnetic separation for optical interference-free detection of atrazine. *Sens. Actuators B Chem.* **2021**, *329*, 129075. [CrossRef]

17. Yin, Y.; Li, Q.; Ma, S.; Liu, H.; Dong, B.; Yang, J.; Liu, D. Prussian Blue as a Highly Sensitive and Background-Free Resonant Raman Reporter. *Anal. Chem.* **2017**, *89*, 1551–1557. [CrossRef]
18. Li, M.; Wang, J.Y.; Chen, Q.Q.; Lin, L.H.; Radjenovic, P.; Zhang, H.; Luo, S.Y.; Tian, Z.Q.; Li, J.F. Background-Free Quantitative Surface Enhanced Raman Spectroscopy Analysis Using Core-Shell Nanoparticles with an Inherent Internal Standard. *Anal. Chem.* **2019**, *91*, 15025–15031. [CrossRef]
19. Gao, X.; Yin, Y.; Wu, H.; Hao, Z.; Li, J.; Wang, S.; Liu, Y. Integrated SERS Platform for Reliable Detection and Photothermal Elimination of Bacteria in Whole Blood Samples. *Anal. Chem.* **2021**, *93*, 1569–1577. [CrossRef]
20. Bi, Y.; Di, H.; Zeng, E.; Li, Q.; Li, W.; Yang, J.; Liu, D. Reliable Quantification of pH Variation in Live Cells Using Prussian Blue-Caged Surface-Enhanced Raman Scattering Probes. *Anal. Chem.* **2020**, *92*, 9574–9582. [CrossRef]
21. Cruz-Aguado, J.A.; Penner, G. Determination of ochratoxin A with a DNA aptamer. *J. Agric. Food Chem.* **2008**, *56*, 10456–10461. [CrossRef] [PubMed]
22. Frens, G. Controlled nucleation for the regulation of the particle size in monodisperse gold suspensions. *Nature* **1973**, *241*, 20–22. [CrossRef]
23. Wang, C.; Dong, X.; Liu, Q.; Wang, K. Label-free colorimetric aptasensor for sensitive detection of ochratoxin A utilizing hybridization chain reaction. *Anal. Chim. Acta* **2015**, *860*, 83–88. [CrossRef]
24. Xia, X.; Wang, Y.; Yang, H.; Dong, Y.; Zhang, K.; Lu, Y.; Deng, R.; He, Q. Enzyme-free amplified and ultrafast detection of aflatoxin B1 using dual-terminal proximity aptamer probes. *Food Chem.* **2019**, *283*, 32–38. [CrossRef] [PubMed]
25. Zhao, X.; Wang, Y.; Li, J.; Huo, B.; Qin, Y.; Zhang, J.; Chen, M.; Peng, Y.; Bai, J.; Li, S. A fluorescence aptasensor based on controlled zirconium–based MOFs for the highly sensitive detection of T–2 toxin. *Spectrochim. Acta Part A* **2021**, *259*, 119893. [CrossRef] [PubMed]
26. Tay, L.L.; Hulse, J.; Kennedy, D.; Pezacki, J.P. Surface-Enhanced Raman and Resonant Rayleigh Scatterings From Adsorbate Saturated Nanoparticles. *J. Phys. Chem. C* **2010**, *114*, 7356–7363. [CrossRef]
27. Kuang, H.; Chen, W.; Xu, D.; Xu, L.; Zhu, Y.; Liu, L.; Chu, H.; Peng, C.; Xu, C.; Zhu, S. Fabricated aptamer-based electrochemical "signal-off" sensor of ochratoxin A. *Biosens. Bioelectron.* **2010**, *26*, 710–716. [CrossRef]
28. Li, A.; Tang, L.; Song, D.; Song, S.; Ma, W.; Xu, L.; Kuang, H.; Wu, X.; Liu, L.; Chen, X. A SERS-active sensor based on heterogeneous gold nanostar core–silver nanoparticle satellite assemblies for ultrasensitive detection of aflatoxin B1. *Nanoscale* **2016**, *8*, 1873–1878. [CrossRef]
29. Kwon, Y.S.; Nguyen, V.T.; Park, J.G.; Gu, M.B. Detection of iprobenfos and edifenphos using a new multi-aptasensor. *Anal. Chim. Acta* **2015**, *868*, 60–66. [CrossRef]
30. Selvolini, G.; Bajan, I.; Hosu, O.; Cristea, C.; Sandulescu, R.; Marrazza, G. DNA-Based Sensor for the Detection of an Organophosphorus Pesticide: Profenofos. *Sensors* **2018**, *18*, 2035. [CrossRef]
31. Zhao, Y.; Wang, Y.; Yang, R.; Zhang, H.; Zhao, Y.; Miao, X.; Lu, L. A zero-background fluorescent aptasensor for ultrasensitive detection of pesticides based on magnetic three-dimensional DNA walker and poly(T) -templated copper nanoparticles. *Sens. Actuators B* **2021**, *343*, 130172. [CrossRef]
32. Fan, L.; Zhao, G.; Shi, H.; Liu, M.; Li, Z. A highly selective electrochemical impedance spectroscopy-based aptasensor for sensitive detection of acetamiprid. *Biosens. Bioelectron.* **2013**, *43*, 12–18. [CrossRef] [PubMed]
33. Luo, Y.; Jin, Z.; Wang, J.; Ding, P.; Pei, R. The isolation of a DNA aptamer to develop a fluorescent aptasensor for the thiamethoxam pesticide. *Analyst* **2021**, *146*, 1986–1995. [CrossRef] [PubMed]
34. Rapini, R.; Cincinelli, A.; Marrazza, G. Acetamiprid multidetection by disposable electrochemical DNA aptasensor. *Talanta* **2016**, *161*, 15–21. [CrossRef]

foods

MDPI

Article

pH-Regulated Strategy and Mechanism of Antibody Orientation on Magnetic Beads for Improving Capture Performance of *Staphylococcus* Species

Fuying Kang [1,†], Yin Yang [2,†], Jingwen Li [1], Erning Chen [1], Tian Hong [1], Lulu Zhao [1] and Meihong Du [1,*]

[1] Institute of Analysis and Testing, Beijing Academy of Science and Technology, Beijing Center of Physical and Chemical Analysis, Beijing 100094, China
[2] Beijing Academy of Science and Technology, Beijing 100094, China
* Correspondence: dumeihong@bcpca.ac.cn; Tel.: +86-010-5871-7271
† These authors have contributed equally to this work.

Highlights:

What are the main findings?
- An pH-regulated strategy can achieve antibody orientation on the surface of magnetic beads.

What is the implication of the main finding?
- The capture efficiency for *Staphylococcus aureus* of immunomagnetic beads prepared at pH 8.0 was improved.
- The antibody orientation mechanism was demonstrated using a quantum dots labeled antigen, antigen-binding fragment (Fab) accessibility assay and lysine mimicking.

Abstract: Immunomagnetic beads (IMBs) have been widely used to capture and isolate target pathogens from complex food samples. The orientation of the antibody immobilized on the surface of magnetic beads (MBs) is closely related to the effective recognition with an antigen. We put forward an available strategy to orient the antibody on the surface of MBs by changing the charged amino group ratio of the reactive amino groups at optimal pH value. Quantum dots labeling antigen assay, antigen-binding fragment (Fab) accessibility assay and lysine mimicking were used for the first time to skillfully illustrate the antibody orientation mechanism. This revealed that the positively charged ε-NH_2 group of lysine on the Fc relative to the uncharged amino terminus on Fab was preferentially adsorbed on the surface of MBs with a negatively charged group at pH 8.0, resulting in antigen binding sites of antibody fully exposed. This study contributes to the understanding of the antibody orientation on the surface of MBs and the potential application of IMBs in the separation and detection of pathogenic bacteria in food samples.

Keywords: antibody orientation; pH regulation; immunomagnetic beads; foodborne pathogen; capture efficiency

Citation: Kang, F.; Yang, Y.; Li, J.; Chen, E.; Hong, T.; Zhao, L.; Du, M. pH-Regulated Strategy and Mechanism of Antibody Orientation on Magnetic Beads for Improving Capture Performance of *Staphylococcus* Species. *Foods* **2022**, *11*, 3599. https://doi.org/10.3390/foods11223599

Academic Editor: Myunghee Kim

Received: 16 September 2022
Accepted: 9 November 2022
Published: 11 November 2022

Publisher's Note: MDPI stays neutral with regard to jurisdictional claims in published maps and institutional affiliations.

Copyright: © 2022 by the authors. Licensee MDPI, Basel, Switzerland. This article is an open access article distributed under the terms and conditions of the Creative Commons Attribution (CC BY) license (https://creativecommons.org/licenses/by/4.0/).

1. Introduction

Magnetic beads (MBs) have been extensively used in the fields of biosensors, biomedicine, and biotechnology due to their unique superparamagnetic properties and basic nanostructured characteristics, such as easy functionalization and larger surface to volume ratio [1–4]. The immunomagnetic beads (IMBs) prepared by immobilizing monoclonal or polyclonal antibodies on the surface of magnetic beads can specifically capture and separate the target and are widely used in the fields of food, hygiene, and environment.

Many studies have reported the potential application of IMBs in the sample pretreatment of the rapid detection of pathogenic bacteria [5–8]. Although the reaction system of IMS for foodborne pathogens, such as the amount of IMBs, and immunoreaction time has been optimized, the capture efficiency of IMBs is still low [2,3,9], and a high capture performance of target bacteria has become the main goal pursued by researchers [10]. It must be said that the capture and separation efficiency of the target bacteria are closely associated with the accessibility of the Fab of the antibody controlled by its orientation and loading capacity [11–15]. However, the most common strategy for antibodies immobilization is to adopt N-hydroxysuccinimide/1-ethyl-3-(3-dimethylamino) propyl carbodiimide hydrochloride (NHS/EDC) chemistry to activate carboxyl groups on the surface of functionalized solid surface for covalently crosslinking with reactive amine groups of the antibody [16–18]. Therefore, covalent coupling strategies do not consider the orientation of the immobilized antibody, and random immobilization of the antibody on solid surface will occur. As a result, the antigen binding sites of the antibody will be blocked and cannot effectively bind the target bacteria, leading to a low capture efficiency.

To date, some effective methods to orient antibodies have been developed, one of which is crosslinking the antibody through carbohydrates groups in the Fc of the antibody with a hydrazine surface [19–21]. Additionally, another option is to introduce protein A or protein G, which can specifically bind to the Fc of the antibody [22–25]. Other strategies involve the use of engineered antibodies by introducing site-specific modifications, such as histidine tags, biotins, and click chemistry reactive groups [26–30]. All of these approaches are based on the immobilization of the antibody through the non-antigen binding Fc region in order to make the Fab more available for antigen recognition. However, the above methods not only affect antibody activity, but also require the recombinant expression of an antibody with further antibody engineering, which is a complex process or involves expensive immobilized proteins.

It is common knowledge that the basic structure of an antibody contains Fab and Fc. Generally, there are more basic amino acids than acidic amino acids at the Fab terminal of the antibody, i.e., there are more amino groups at the Fab terminal. However, at the Fc terminal of the antibody, the quantity distribution of the two amino acids is more uniform, so the isoelectric point of the Fab terminal is slightly higher than that of the Fc terminal. Such different pH values will lead to an inconsistent charge distribution of antibodies, which will affect the coupling effect of antibodies. In a recent report, the surface electrical properties of the nanomaterial and basic characteristics of the antibody, such as charge distribution and hydrophilicity were considered to develop a simple universal method to orient the antibody [31], suggesting that the pH value of the reaction solution can affect the behavior of an antibody and potentially modulate the directional fixation of antibodies on the surface of nanomaterials [15,32]. Some novel strategies have been proposed: an antibody is first adsorbed into the nanomaterial surface through non-covalent interactions before being coupled to the surface [33,34], which is a process easily controlled by reaction conditions. Other researchers applied an external electric field to orient antibodies on two-dimensional surfaces achieving more than a 100% improvement in the signal-to-noise ratio [35], and a weak electric field in the surrounding environment can affect the interaction between antibodies and surface-charged nanoparticles. Some studies used a model IgG protein (PDB: 1IGY) to investigate the orientation mechanism of antibodies on gold nanoparticles, such as surface amino acid charge distribution and the amount of acid or alkaline amino acid in the Fab or Fc [33,36]. Because different antibody proteins contain different amino acid sequence structures, and each amino acid has its own isoelectric point, which means that different antibodies have different isoelectric points, a slight sequence variation may alter the structure and electrical properties of the antibody surface [37–39], thus immobilizing different antibodies on nanomaterials at different pH values (pH 5–8.5), which may lead to different orientations [40,41].

Here, we extend the methodology to the orientation of Abs on the surface of carboxylated magnetic beads and its new application fields in food microbiological detection. Based on the amino terminus and lysine side chain amino groups with significantly different pKa values [15], we propose a strategy to control antibody orientation on MBs by modulating the degree of ionization of reactive amino groups. Firstly, the amino terminal of the Fab (pKa = 7.5) and the amino group from the lysine side chain on the Fc (pKa = 10.0) [15] could be regulated with charged properties to achieve the orientation of the Fc, and then N-hydroxysulfosuccinimide (sulfo-NHS) was used to obtain the negatively charged surface of the carboxyl MBs based on the NHS/EDC crosslinking chemistry (Figure 1).

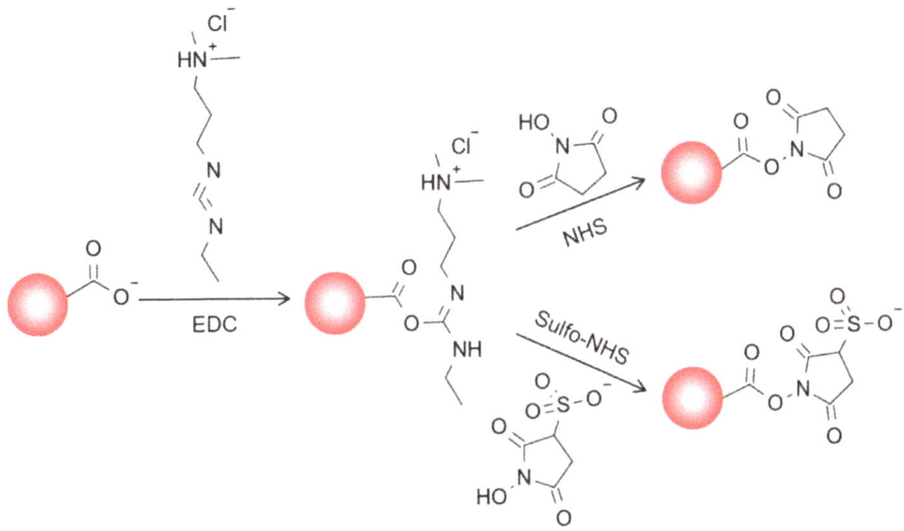

Figure 1. Scheme of activation mechanism of carboxyl MBs based on NHS/EDC chemistry.

Considering the prevalence of food contaminated by *Staphylococcus*, we chose the anti-*Staphylococcus* antibody as a representative to prepare the IMBs to validate the feasibility of the strategy. Meanwhile, to probe the mechanism of antibody orientation, we employed a quantum dots labeling antigen, Fab accessibility assay and lysine mimicking to indirectly characterize the behavior of antibodies on the surface of MBs.

2. Methods/Experimental Section

2.1. Reagents, Materials, and Apparatus

Carboxyl magnetic beads and preservation solution were obtained from BioMag Scientific Inc. (Wuxi, China), CdSe/ZnS quantum dots (QDs) were purchased from Xingzi New Material Technology Development (Shanghai, China), the *Staphylococcus* monoclonal antibody was obtained from LSBio Inc. (Seattle, WA, USA), goat anti-mouse IgG Fc (anti-Fc) was obtained from ImmunoChemistry Technologies, LLC, (Bloomington, MN, USA), and *Staphylococcus* enterotoxin B (SEB) was purchased from Toxin technology, Inc. (Sarasota, FL, USA). The Micro Bicinchoninic Acid (BCA) Protein Assay Kit, Varioskan Flash Microplate Reader, and DynaMag magnetic separator were purchased from Thermo Fisher Scientific Inc. (Waltham, MA, USA), and the Amino Acid Assay Kit and 2-morpholinoethanesulfonic acid monohydrate (MES) were obtained from Solarbio Science & Technology Co., Ltd. (Beijing, China). Ultra-high temperature instantaneous sterilization milk was obtained from a local supermarket; 1-ethyl-3-(3-dimethylamino) propyl carbodiimide hydrochloride crystalline (EDC), N-hydroxysulfosuccinimide sodium salt (sulfo-NHS), boric acid, sodium tetraborate, N-tert-butyloxycarbonyl(BOC)-L-lysine(ε-NH_2-lys), and Nε-BOC-L-lysine(α-NH_2-lys) were purchased from J&K (Beijing, China); Nutrient Broth (NB) and plate count agar (PCA) were purchased from Land Bridge Technology Co., Ltd. (Beijing, China). *S. aureus* (ATCC 25923) was obtained from American Type Culture Collection (ATCC, Manassas, VA, USA), the Zeiss Fluorescence Microscope was purchased from Carl Zeiss Microscopy GmbH (Jena, Germany), QB-628 Rolling Incubator was purchased from Haimen-Kylin-Bell Lab Instruments Co., Ltd. (Nantong, China), and the Zetasizer nano ZS was purchased from Malvern Panalytical Ltd. (Malvern, UK).

2.2. Preparation of IMBs

2.2.1. Activation of Carboxyl MBs

Carboxyl MBs (1 mg) were added to a 2 mL low-affinity microcentrifuge tube, washed twice with 1 mL of MEST buffer (0.1 M MES pH 6.0, 0.05% Tween20), and incubated with 10 mg EDC mixed with 10 mg sulfo-NHS (dissolved in MEST buffer) at 25 °C for 30 min. The mixture was then separated by a magnetic separator for 1 min. The supernatant was discarded, and the activated carboxyl magnetic beads (aMBs) were used immediately.

2.2.2. Antibody Immobilization on MBs

The aMBs (1 mg) were resuspended with 200 μg of anti-*Staphylococcus* antibody [7], which was excessive compared to the previous study in this laboratory, and was dissolved in 0.05 M phosphate-buffered saline containing 0.05% Tween20 (PBST) (pH 6.0 and 7.0) or 0.05 M borate buffer (pH 8.0, 0.05% Tween20) in advance, and then incubated for 2 h at room temperature on a vertical rotating mixer. The activation sites on MBs were blocked by incubation with bovine serum albumin (BSA) for 1 h. The prepared IMBs were washed twice with PBST and stored in preservation solution at 4 °C.

2.3. Characteristics of IMBs

2.3.1. Antibody Binding Quantification

The amount of antibody was quantified by measuring the protein concentration with the bicinchoninic acid (BCA) protein quantitation kit. Then, 150 μL of the sample and a 150 μL aliquot of the working reagent were thoroughly mixed and incubated for 2 h at 37 °C. Absorbance at 562 nm was measured with a microplate reader to determine the protein content based on a standard curve. The amount of antibody bound on the surface of the MBs was calculated as the protein concentration difference before and after binding to the MBs.

2.3.2. Size Distribution and Dispersity Characteristics

IMBs and aMBs were prepared at a concentration of 0.33 mg/mL in PBST or different pH buffers, and dispersed by sonication to evaluate the size distribution and zeta potential using the Zetasizer nano ZS instrument. The dispersed states of MBs and IMBs at the same concentration were observed by a fluorescence microscope with 40× objective lens. The anti-*Staphylococcus* antibody was diluted to a final concentration of 60 μg/mL in PBS to measure size distribution.

2.3.3. Capture Efficiency

One hundred micrograms of IMBs was incubated with 10^4 colony-forming unit (CFU) of *S. aureus* in 1 mL PBST or milk, shaken on a vertical rotating mixer at room temperature for 1 h, and then separated by a magnetic separator. The IMB–bacterial complexes and supernatant were separately diluted to appropriate concentrations and cultured on PCA for 24 h to count the number of bacterial colonies formed. The capture efficiency of the IMBs was equal to the percentage of bacteria captured by IMBs relative to the total bacterial count.

2.3.4. Measurement of the Maximum Binding Capacity

The *S. aureus* suspension cultured overnight at 37 °C in NB was diluted in 1 mL PBST to final concentrations of 10^2, 10^3, 10^4, 10^5, 10^6 CFU/mL, and each diluted suspension was incubated with 100 μg of IMBs at room temperature for 1 h, and then separated by a magnetic separator. The capture efficiency of the IMBs was determined by the aforementioned method.

All used reagents, equipment and waste generated in the experiment were sterilized by autoclaving at 120 °C for 1 h to prevent the spread of pathogens.

2.4. Mechanism of Antibody Orientation

2.4.1. Quantum Dots Labeling Antigen Assay

A QDs stock solution (80 μL, 5 mg/mL) was activated with 4 mg EDC and 2 mg sulfo-NHS for 30 min, followed by the addition of 50 μg SEB, and then incubated in the dark for

3 h. The unreacted SEB protein was removed from the system by cryogenic centrifugation and washed twice with PBST. QDs and SEB conjugates were collected and resuspended in 500 μL PBST.

IMBs (50 μg) and the above functionalized QDs (20 μL) were incubated in the dark for 30 min, and then separated by a magnetic separator. The IMB-QD complexes were washed twice with PBST and resuspended in PBST. The obtained supernatant and IMB-QD complexes were used in the subsequent fluorescence analysis.

Confocal fluorescence microscopy imaging was performed on a confocal laser scanning fluorescence microscope, and the IMB-QD complexes were prepared by the aforementioned protocol at the same concentration and measured using the same parameters (with 405 nm of laser and 2 μs of exposure time).

The fluorescence intensity values of the above supernatant and IMB-QD complexes were measured on a microplate reader from 400 to 600 nm with an excitation wavelength of 340 nm.

2.4.2. Fab Accessibility Assay

A mixture of 100 μg IMBs and 10 μg anti-Fc in 200 μL PBS was incubated at 25 °C for 1 h, and the supernatant was collected after separation by a magnetic separator. The amount of anti-Fc was quantified using a BCA protein quantitation kit. The amount of anti-Fc bound to the IMBs was calculated by determining the protein concentration difference between the initial concentration and the concentration in the supernatant according to a standard curve.

2.4.3. Crosslinking Ratio Analysis of ε-NH_2-lys and α-NH_2-lys on the Surface of the MBs

The aMBs were incubated with ε-NH_2-lys and α-NH_2-lys diluted with 0.05 M PBS (pH 6.0 or 7.0) for 2 h, respectively. The crosslinking ratio was calculated based on the amino acid content before and after crosslinking with MBs. The amino acid content was quantified using an amino acid assay kit based on the principle that the amino group of amino acids can react with indigohydrone to produce blue–violet compounds and measured on a microplate reader at 570 nm.

3. Results and Discussion

3.1. Characterization of IMBs Properties

To investigate the effect of pH condition on antibody adsorption and orientation, the antibody was pre-ionized at pH 6.0, 7.0, and 8.0, and coupled with MBs to produce pH6-IMBs, pH7-IMBs and pH8-IMBs. A BCA assay was employed to analyze the antibody binding content. The amounts of antibodies bound on the surface of MBs at pH 6.0 and pH 7.0 were higher than the amount of antibody at pH 8.0 (Figure 2A), which is consistent with the recommended lower pH in NHS/EDC coupling protocol. We speculated that the surface of negatively charged carboxyl magnetic beads would adsorb more antibodies because activated amino groups on the antibodies were positively charged below pH 7.5. Furthermore, we analyzed the size distribution of the antibody, the MBs, and the IMBs. The observed size distribution of IMBs was larger than that of the sum of the MB and antibody size due to the presence of monomers (16 nm) and polymers (79 nm) of the *Staphylococcus* antibody (Figure 2B,C). Finally, the dispersity of the aMBs and IMBs was characterized by microscopy imaging and the zeta potential measurement, taking into account its influence on the antibody binding capacity and target recognition [42]. The results show that MBs exhibited a better dispersion and were more negatively charged compared to the IMBs due to the antibody coating and blocking with BSA [4], while there was no significant difference between IMBs at pH 6.0–8.0 (Figure 3, Table 1). Based on the above data of the binding amount and dispersion of the IMBs, the higher capturing capacity of pH6-IMBs, pH7-IMBs and pH8-IMBs on the target antigen would ideally be taken for granted, following to verify this inference via a capture efficiency measurement.

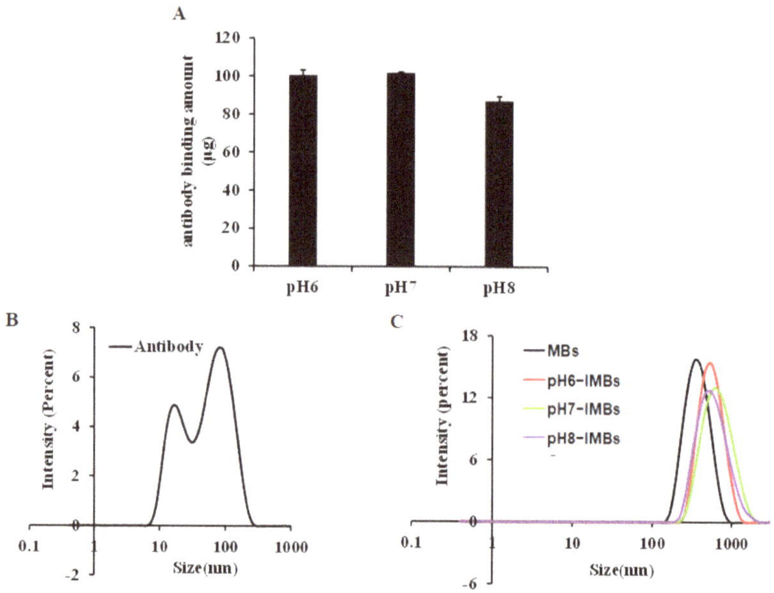

Figure 2. The characterization of IMBs properties. (**A**) Amount of antibody binding to MBs at pH 6.0, 7.0 and 8.0, (**B**) the size distribution assays of antibody, (**C**) and the size distribution assays of MBs, pH6−IMBs, pH7−IMBs, and pH8−IMBs.

Figure 3. Microscopy images of MBs (**A**) pH6-IMBs (**B**) pH7-IMBs (**C**) and pH8-IMBs (**D**).

Table 1. The dispersity of aMBs and IMBs.

Sample	Zeta Potential (mV)
aMBs-pH6	−47.73
aMBs-pH7	−41.87
aMBs-pH8	−57.07
pH6-IMBs	−15.33
pH7-IMBs	−17.23
pH8-IMBs	−15.83

3.2. Capture Efficiency of pH6-IMBs, pH7-IMBs and pH8-IMBs

The capture efficiency on *S. aureus* of pH6-IMBs, pH7-IMBs and pH8-IMBs with equal amounts was measured in PBS or milk. The results show that the capture efficiency of pH8-IMBs was higher than that of pH6-IMBs or pH7-IMBs (Figure 4A). The maximum capacity of the binding antigen of pH6-IMBs, pH7-IMBs and pH8-IMBs was 10^5 CFU (Figure 4B); meanwhile, at 10^3–10^6 CFU/mL of *S. aureus*, pH8-IMBs have certain advantages in capture efficiency compared to pH6-IMBs and pH7-IMBs. However, at 10^2 CFU/mL of the *S. aureus*, the capture efficiency of pH8-IMBs was not significantly higher than that of pH6-IMBs and pH7-IMBs as the amount of used IMBs exceeded the actual amount needed. Furthermore, we investigated the capture efficiency of pH6-IMBs pH7-IMBs and pH8-IMBs with different amounts on 10^0 CFU/mL of *S. aureus*. The results show that the capture efficiency of all the IMBs was 100% when 20 or 50 µg of IMBs were used, indicating that the amount of used IMBs was excessive. While the amount of IMBs was reduced to 10 µg, the capture efficiency on target bacteria of pH8-IMBs was significantly higher than that of pH6-IMBs or pH7-IMBs (Figure 4C,D). The above results show that the antibody binding capacity of pH8-IMBs was lower, while the capture efficiency was higher than that of pH6-IMBs or pH7-IMBs, and these differences were not related to the dispersity. We speculate that the Fab fragment that contains the antigen binding site was in a different exposure status at pH 6.0, 7.0, and 8.0 during antibody immobilization, and the more the Fab fragment was exposed, the more the antigen was recognized.

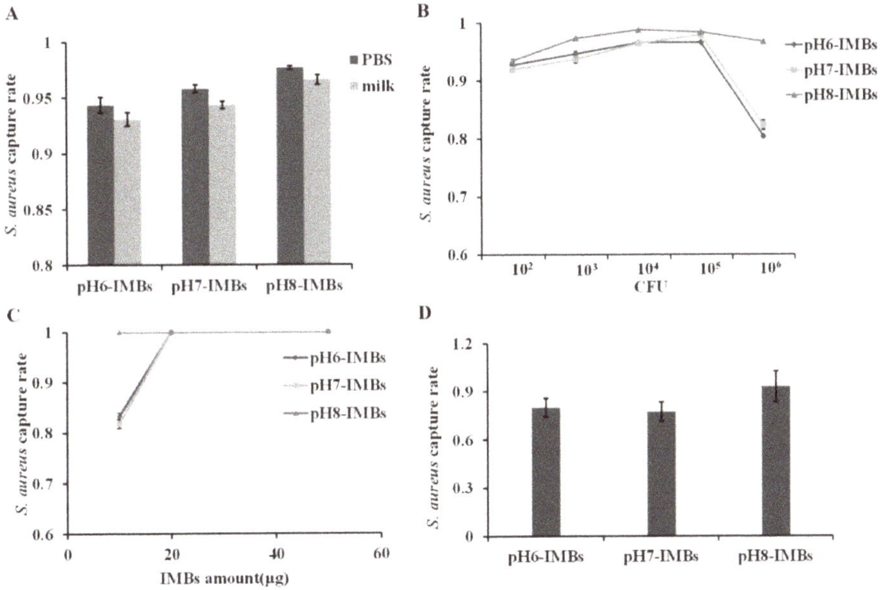

Figure 4. Capture efficiency on *S. aureus* of pH6-IMBs, pH7-IMBs, and pH8-IMBs. (**A**) The capture

efficiency of 100 μg IMBs on *S. aureus* in PBS and in milk, (**B**) Maximum binding capacity analysis of 100 μg IMBs, (**C**) The capture efficiency of IMBs with 10, 20 and 50 μg on 10^0 CFU/mL of *S. aureus*, (**D**) The capture efficiency of 10 μg of IMBs on 10^0 CFU/mL of *S. aureus*.

3.3. Fluorescence Analysis of Antibody Orientation on MBs

To visualize target antigen captured and determine the orientation of the antibody on the surface of MBs, we taken an approach: antigen SEB labeled by QD (QD-SEB) bound to the surface of IMBs to form the IMB-QD complexes by a further immune response step, and the fluorescence intensity of IMB-QDs were analyzed after magnetic separation, which should be related to the amount of antigen captured and exposure status of the Fab fragment (Figure 5A). First, we measured the excitation and emission spectrum of QD-SEB conjugates and QDs by microplate reader. After coupling with the SEB, the maximum excitation and emission wavelength of QD-SEB complexes was slightly red-shifted compared to QDs due to the surrounding organic layer (antigen) [43] (Figure 5E). Additionally, the fluorescence microscopy imaging visually indicated that pH6-IMBs, pH7-IMBs and pH8-IMBs can recognize and capture QD-SEB to form IMB-QD complexes (Figure 5B–D). The fluorescence intensity in the supernatant (pH6-S-QD, pH7-S-QD and pH8-S-QD) was from the unbound QD-SEB complexes after pH6-IMBs, pH7-IMBs and pH8-IMBs capture and separation. These results show that the fluorescence intensity of pH8-IMB-QD complexes is higher than pH6-IMB-QD or pH7-IMB-QD complexes, and that of the corresponding supernatant is lower than that of pH6-S-QD and pH7-S-QD. The above experiments confirmed that pH8-IMBs possessed more exposed Fab fragments and more antigen recognition sites to bind more antigen molecules than pH6-IMBs and pH7-IMBs (Figure 5F,G).

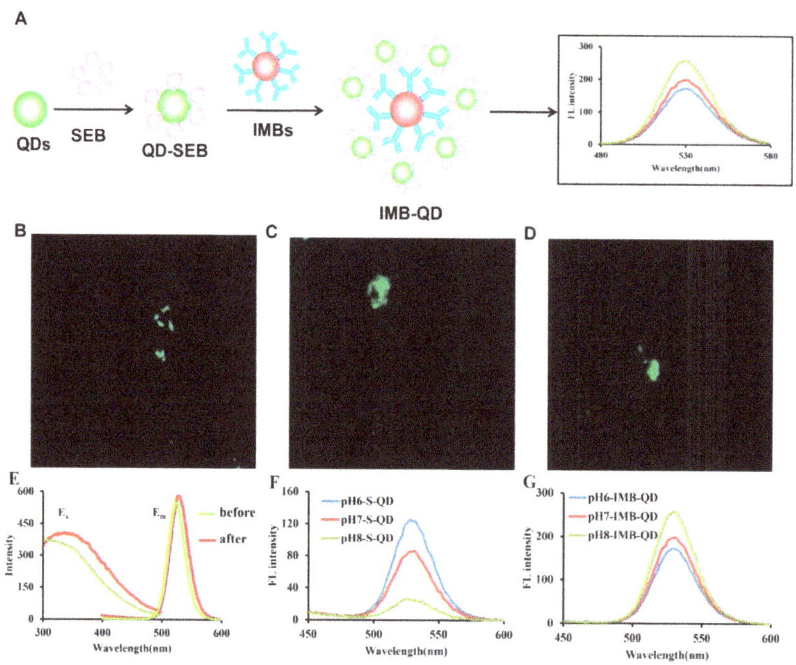

Figure 5. Fluorescence analysis of antibody orientation on MBs. (**A**) Schematic represent for fluorescence analysis, fluorescence microscopy images of pH6-IMB-QD (**B**), pH7-IMB-QD (**C**) and pH8-IMB-QD (**D**), the excitation and emission spectra of QDs and QD-SEB complexes (**E**), fluorescence (FL) intensity analysis of QDs in the supernatant (**F**) and IMB-QD complexes (**G**).

3.4. Fab Accessibility Analysis on IMBs

In order to obtain more direct evidence that the Fab fragment of the pH8-IMBs is more exposed than in the pH6-IMBs or pH7-IMBs, and based on the fact that anti-Fc antibody can specifically recognize the exposed Fc region of the IMBs, we evaluated the accessibility of Fab on IMBs by determining the amount of anti-Fc antibody bound to IMBs. Figure 6A reveals that higher pH values would result in fewer anti-Fc antibodies bound to IMBs, indicating that the Fc region of the *Staphylococcus* antibody was more attached to the surface of the MBs at pH 8.0. In other words, the Fab fragment of pH8-IMBs was more exposed to the antigen, which increased the likelihood of IMBs of recognizing *S. aureus*, resulting in higher capture efficiency.

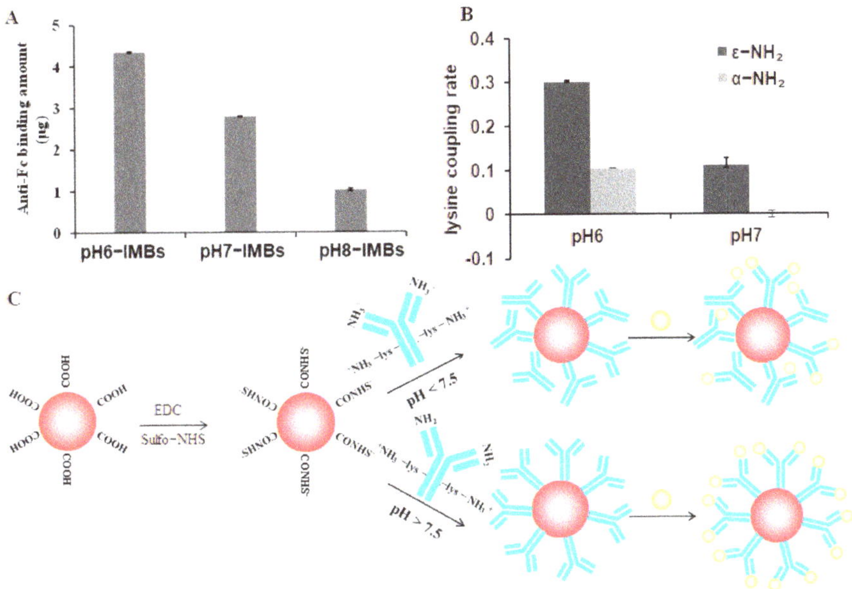

Figure 6. Mechanism of antibody orientation. (**A**) Amount of anti−Fc antibody bound to IMBs, (**B**) The coupling ratio assay of α−NH_2−lys and ε−NH_2−lys at pH 6.0 and 7.0, (**C**) The proposed mechanism of antibody orientation on MBs for *S. aureus*. Red spheres represent MBs, yellow spheres represent *S. aureus*, and blue stick structures represent antibody.

3.5. Lysine Mimicking

Considering the complex microenvironment in which lysine is located, we only chose the representative ε-NH_2-lys (α-NH_2 is protected by a BOC protecting group) and α-NH_2-lys (ε-NH_2 is protected by a BOC protecting group) to mimic the ε-NH_2 of lysine and the amino terminal of the antibody, respectively, in order to accurately understand the mechanism of the oriented immobilization of antibody. We studied the coupling ratio of aMBs with α-NH_2-lys and ε-NH_2-lys at pH 6.0, 7.0 and 8.0. Although the data of the coupling rate were not obtained due to the poor solubility of amino acids at pH 8.0, the results also show that the coupling ratio of α-NH_2-lys significantly decreased compared to that of ε-NH_2-lys at a higher pH value (Figure 6B). The positively charged α-NH_2-lys and ε-NH_2-lys could couple with negatively charged aMBs at pH 6.0, whereas only the positively charged ε-NH_2-lys could couple with aMBs at pH 7.0. Similarly, at pH 8.0, the ε-NH_2 group of lysine residues in the Fc region was more prone to being adsorbed in aMBs relative to the amino terminus of the antibody, which contributed to the oriented immobilization of the Fc fragment on the surface of MBs, thereby fully exposing the Fab fragment and enabling the recognition of more antigens.

4. Conclusions

In this study, we proposed an effective approach to control antibody orientation on the surface of MBs by modulating the degree of ionization of reactive amino groups. The mechanism of oriented immobilization of antibody on MBs was studied through the use of a QDs labeled antigen, Fab accessibility assay and lysine mimicking. MBs activated by EDC/sulfo-NHS were negatively charged and adsorbed the positively charged amino groups of the antibody (amino terminus and ε-NH_2 of lysine) through electrostatic interactions before crosslinking with the MBs. The amino terminus of the Fab fragment and the ε-NH_2 of lysine from Fc region were both positively charged at pH < 7.5, leading to the random immobilization of the antibody on MBs. At pH > 7.5, the positively charged ε-NH_2 group of lysine was preferentially adsorbed on the negative surface relative to the uncharged amino terminus, resulting in Fc-oriented immobilization and a more exposed Fab fragment (Figure 6C). As a result, we improved the capture efficiency of IMBs on *S. aureus* in PBS or milk by controlling the orientation of *Staphylococcus* antibodies immobilized on MBs at pH 8.0, and higher capture efficiencies can be achieved with a lower amount of pH8-IMBs. This strategy would be very useful for the preparation of IMBs for effectively capturing and separating pathogenic bacteria from a complicated food matrix with lower cost, thereby improving specificity and sensitivity integrated with PCR, enzyme-linked immunosorbent assay (ELISA), luminescence and electrochemistry for the detection of trace pathogenic microorganisms.

Author Contributions: M.D. conceived the study and revised the manuscript. F.K. and Y.Y. carried out experiments, analyzed data and wrote the paper. J.L., E.C., T.H. and L.Z. assisted in the experiment and result analysis. All authors have read and agreed to the published version of the manuscript.

Funding: This work has been funded by the National Natural Science Foundation of China [grant number 31972146], High-level Innovation Team Program (grant number HIT202201) and BJAST Budding Talent Program (grant number BGS202116).

Data Availability Statement: All data and materials during the current study are available from the corresponding author on reasonable request.

Conflicts of Interest: The authors declare no conflict of interest.

Abbreviations

IMBs	immunomagnetic beads
QDs	quantum dots
EDC	N-(3-dimethylaminopropyl)-N′-ethylcarbodiimide hydrochloride crystalline
	Sulfo-NHS: N-hydroxysulfosuccinimide sodium salt
BCA	bicinchoninic acid
BSA	bovine serum albumin
PCA	plate count agar
NB	nutrient broth
SEB	Staphylococcus enterotoxin B
BOC	tert-butyloxycarbonyl
CFU	colony-forming unit
Fab	antigen-binding fragment
Fc	fragment crystallizable

References

1. Nam, K.C.; Han, Y.S.; Lee, J.-M.; Kim, S.C.; Cho, G.; Park, B.J. Photo-Functionalized Magnetic Nanoparticles as a Nanocarrier of Photodynamic Anticancer Agent for Biomedical Theragnostics. *Cancers* **2020**, *12*, 571. [CrossRef] [PubMed]
2. Kyeong, S.; Kim, J.; Chang, H.; Lee, S.H.; Son, B.S.; Lee, J.H.; Rho, W.Y.; Pham, X.H.; Jun, B.H. Magnetic Nanoparticles. *Adv. Exp. Med. Biol.* **2021**, *1309*, 191–215. [PubMed]
3. Pastucha, M.; Farka, Z.; Lacina, K.; Mikušová, Z.; Skládal, P. Magnetic nanoparticles for smart electrochemical immunoassays: A review on recent developments. *Mikrochim. Acta* **2019**, *186*, 312. [CrossRef] [PubMed]

4. Sharma, S.; Ram, P. Capturing of Magnetic Nanoparticles in a Fluidic Channel for Magnetic Drug Targeting. *J. Nanosci. Nanotechnol.* **2021**, *21*, 3588–3595. [CrossRef]
5. Kim, J.-H.; Oh, S.-W. Pretreatment methods for nucleic acid-based rapid detection of pathogens in food: A review. *Food Control* **2021**, *121*, 107575. [CrossRef]
6. Du, M.; Li, J.; Liu, Q.; Wang, Y.; Chen, E.; Kang, F.; Tu, C. Rapid detection of trace Salmonella in milk using an effective pretreatment combined with droplet digital polymerase chain reaction. *Microbiol. Res.* **2021**, *251*, 126838. [CrossRef]
7. Du, M.; Li, J.; Zhao, R.; Yang, Y.; Wang, Y.; Ma, K.; Cheng, X.; Wan, Y.; Wu, X. Effective pre-treatment technique based on immune-magnetic separation for rapid detection of trace levels of Salmonella in milk. *Food Control* **2018**, *91*, 92–99. [CrossRef]
8. Johari-Ahar, M.; Rashidi, M.R.; Barar, J.; Aghaie, M.; Mohammadnejad, D.; Ramazani, A.; Karami, P.; Coukos, G.; Omidi, Y. An ultra-sensitive impedimetric immunosensor for detection of the serum oncomarker CA-125 in ovarian cancer patients. *Nanoscale* **2015**, *7*, 3768–3779. [CrossRef]
9. Wei, S.; Park, B.-J.; Seo, K.-H.; Oh, D.-H. Highly efficient and specific separation of Staphylococcus aureus from lettuce and milk using Dynabeads protein G conjugates. *Food Sci. Biotechnol.* **2016**, *25*, 1501–1505. [CrossRef]
10. Luciani, M.; Di Febo, T.; Zilli, K.; Di Giannatale, E.; Armillotta, G.; Manna, L.; Minelli, F.; Tittarelli, M.; Caprioli, A. Rapid Detection and Isolation of Escherichia coli O104:H4 from Milk Using Monoclonal Antibody-coated Magnetic Beads. *Front. Microbiol.* **2016**, *15*, 942. [CrossRef]
11. Kausaite-Minkstimiene, A.; Ramanaviciene, A.; Kirlyte, J.; Ramanavicius, A. Comparative Study of Random and Oriented Antibody Immobilization Techniques on the Binding Capacity of Immunosensor. *Anal. Chem.* **2010**, *82*, 6401–6408. [CrossRef] [PubMed]
12. Lin, P.C.; Chen, S.H.; Wang, K.Y.; Chen, M.L.; Adak, A.K.; Hwu, J.R.; Chen, Y.J.; Lin, C.C. Fabrication of Oriented Antibody-Conjugated Magnetic Nanoprobes and Their Immunoaffinity Application. *Anal. Chem.* **2009**, *81*, 8774–8782. [CrossRef] [PubMed]
13. Puertas, S.; Moros, M.; Fernández-Pacheco, R.; Ibarra, M.R.; Grazú, V.; De La Fuente, J.M. Designing novel nano-immunoassays: Antibody orientation versus sensitivity. *J. Phys. D-Appl. Phys.* **2010**, *43*, 474012. [CrossRef]
14. Saha, B.; Songe, P.; Evers, T.H.; Prins, M.W.J. The influence of covalent immobilization conditions on antibody accessibility on nanoparticles. *Analyst* **2017**, *142*, 4247–4256. [CrossRef] [PubMed]
15. Puertas, S.; Villa, M.D.G.; Mendoza, E.; Jiménez-Jorquera, C.; de la Fuente, J.M.; Fernández-Sánchez, C.; Grazú, V. Improving immunosensor performance through oriented immobilization of antibodies on carbon nanotube composite surfaces. *Biosens. Bioelectron.* **2013**, *43*, 274–280. [CrossRef]
16. Song, H.Y.; Zhou, X.; Hobley, J.; Su, X. Comparative Study of Random and Oriented Antibody Immobilization as Measured by Dual Polarization Interferometry and Surface Plasnnon Resonance Spectroscopy. *Langmuir* **2012**, *28*, 997–1004. [CrossRef]
17. Raghav, R.; Srivastava, S. Immobilization strategy for enhancing sensitivity of immunosensors: L-Asparagine-AuNPs as a promising alternative of EDC-NHS activated citrate-AuNPs for antibody immobilization. *Biosens. Bioelectron.* **2016**, *78*, 396–403. [CrossRef]
18. Yan, Q.; Zheng, H.-N.; Jiang, C.; Li, K.; Xiao, S.-J. EDC/NHS activation mechanism of polymethacrylic acid: Anhydride versus NHS-ester. *RSC Adv.* **2015**, *5*, 69939–69947. [CrossRef]
19. Duval, F.; van Beek, T.A.; Zuilhof, H. Key steps towards the oriented immobilization of antibodies using boronic acids. *Analyst* **2015**, *140*, 6467–6472. [CrossRef]
20. Kumar, S.; Aaron, J.; Sokolov, K. Directional conjugation of antibodies to nanoparticles for synthesis of multiplexed optical contrast agents with both delivery and targeting moieties. *Nat. Protoc.* **2008**, *3*, 314–320. [CrossRef]
21. Prieto-Simon, B.; Saint, C.; Voelcker, N.H. Electrochemical Biosensors Featuring Oriented Antibody Immobilization via Electro-grafted and Self-Assembled Hydrazide Chemistry. *Anal. Chem.* **2014**, *86*, 1422–1429. [CrossRef] [PubMed]
22. Seo, J.S.; Lee, S.; Poulter, C.D. Regioselective Covalent Immobilization of Recombinant Antibody-Binding Proteins A, G, and L for Construction of Antibody Arrays. *J. Am. Chem. Soc.* **2013**, *135*, 8973–8980. [CrossRef] [PubMed]
23. Moon, J.; Byun, J.; Kim, H.; Jeong, J.; Lim, E.; Jung, J.; Cho, S.; Cho, W.K.; Kang, T. Surface-Independent and Oriented Immobilization of Antibody via One-Step Polydopamine/Protein G Coating: Application to Influenza Virus Immunoassay. *Macromol. Biosci.* **2019**, *19*, e1800486. [CrossRef] [PubMed]
24. Kaittanis, C.; Santra, S.; Perez, J.M. Role of Nanoparticle Valency in the Nondestructive Magnetic-Relaxation-Mediated Detection and Magnetic Isolation of Cells in Complex Media. *J. Am. Chem. Soc.* **2009**, *131*, 12780–12791. [CrossRef] [PubMed]
25. Granade, T.C.; Workman, S.; Wells, S.K.; Holder, A.N.; Owen, S.M.; Pau, C.-P. Rapid Detection and Differentiation of Antibodies to HIV-1 and HIV-2 Using Multivalent Antigens and Magnetic Immunochromatography Testing. *Clin. Vaccine Immunol.* **2010**, *17*, 1034–1039. [CrossRef]
26. Baio, G.; Fabbi, M.; Salvi, S.; De Totero, D.; Truini, M.; Ferrini, S.; Neumaier, C.E. Two-Step In Vivo Tumor Targeting by Biotin-Conjugated Antibodies and Superparamagnetic Nanoparticles Assessed by Magnetic Resonance Imaging at 1.5 T. *Mol. Imaging Biol.* **2010**, *12*, 305–315. [CrossRef]
27. Zhao, X.; Pan, F.; Garcia-Gancedo, L.; Flewitt, A.J.; Ashley, G.M.; Luo, J.; Lu, J.R. Interfacial recognition of human prostate-specific antigen by immobilized monoclonal antibody: Effects of solution conditions and surface chemistry. *J. R. Soc. Interface* **2012**, *9*, 2457–2467. [CrossRef]
28. Wang, A.; Perera, Y.R.; Davidson, M.B.; Fitzkee, N.C. Electrostatic Interactions and Protein Competition Reveal a Dynamic Surface in Gold Nanoparticle-Protein Adsorption. *J. Phys. Chem. C Nanomater. Interfaces* **2016**, *120*, 24231–24239. [CrossRef]

29. Lin, W.; Insley, T.; Tuttle, M.D.; Zhu, L.; Berthold, D.A.; Král, P.; Rienstra, C.M.; Murphy, C.J. Control of protein orientation on gold nanoparticles. *J. Phys. Chem. C Nanomater. Interfaces* **2015**, *119*, 21035–21043. [CrossRef]
30. Siriwardana, K.; Wang, A.; Vangala, K.; Fitzkee, N.; Zhang, D. Probing the effects of cysteine residues on protein adsorption onto gold nanoparticles using wild-type and mutated GB3 proteins. *Langmuir* **2013**, *29*, 10990–10996. [CrossRef]
31. Wiseman, M.E.; Frank, C.W. Antibody adsorption and orientation on hydrophobic surfaces. *Langmuir* **2012**, *28*, 1765–1774. [CrossRef] [PubMed]
32. Ruiz, G.; Tripathi, K.; Okyem, S.; Driskell, J.D. pH Impacts the Orientation of Antibody Adsorbed onto Gold Nanoparticles. *Bioconjugate Chem.* **2019**, *30*, 1182–1191. [CrossRef] [PubMed]
33. Zhou, J.; Tsao, H.-K.; Sheng, Y.-J.; Jiang, S. Monte Carlo simulations of antibody adsorption and orientation on charged surfaces. *J. Chem. Phys.* **2004**, *121*, 1050–1057. [CrossRef] [PubMed]
34. Chen, S.; Liu, L.; Zhou, J.; Jiang, S. Controlling antibody orientation on charged self-assembled monolayers. *Langmuir* **2003**, *19*, 2859–2864. [CrossRef]
35. Emaminejad, S.; Javanmard, M.; Gupta, C.; Chang, S.; Davis, R.W.; Howe, R.T. Tunable control of antibody immobilization using electric field. *Proc. Natl. Acad. Sci. USA* **2015**, *112*, 1995–1999. [CrossRef]
36. Lou, D.; Ji, L.; Fan, L.; Ji, Y.; Gu, N.; Zhang, Y. Antibody-Oriented Strategy and Mechanism for the Preparation of Fluorescent Nanoprobes for Fast and Sensitive Immunodetection. *Langmuir* **2019**, *35*, 4860–4867. [CrossRef]
37. Eriksson, A.E.; Baase, W.A.; Zhang, X.J.; Heinz, D.W.; Blaber, M.; Baldwin, E.P.; Matthews, B.W. Response of a Protein-Structure to Cavity-Creating Mutations and Its Relation to the Hydrophobic Effect. *Science* **1992**, *255*, 178–183. [CrossRef]
38. Garcia-Seisdedos, H.; Ibarra-Molero, B.; Sanchez-Ruiz, J.M. How many ionizable groups can sit on a protein hydrophobic core? *Proteins-Struct. Funct. Bioinform.* **2012**, *80*, 1–7. [CrossRef]
39. Xu, J.; Baase, W.A.; Baldwin, E.; Matthews, B.W. The response of T4 lysozyme to large-to-small substitutions within the core and its relation to the hydrophobic effect. *Protein Sci.* **1998**, *7*, 158–177. [CrossRef]
40. Rao, S.V.; Anderson, K.W.; Bachas, L.G. Oriented Immobilization of Proteins. *Protein Sci.* **1998**, *128*, 127–143. [CrossRef]
41. Fuentes, M.; Mateo, C.; Guisán, J.; Fernández-Lafuente, R. Preparation of Inert Magnetic Nanoparticles for the Directed Immobilization of Antibodies. *Biosens. Bioelectron.* **2005**, *20*, 1380–1387. [CrossRef] [PubMed]
42. Sun, J.; He, M.; Liu, X.; Gu, N. Optimizing colloidal dispersity of magnetic nanoparticles based on magnetic separation with magnetic nanowires array. *Appl. Phys. A-Mater. Sci. Process.* **2015**, *118*, 569–577. [CrossRef]
43. Zhao, Y.; Ye, M.; Chao, Q.; Jia, N.; Ge, Y.; Shen, H. Simultaneous Detection of Multifood-Borne Pathogenic Bacteria Based on Functionalized Quantum Dots Coupled with Immunomagnetic Separation in Food Samples. *J. Agric. Food Chem.* **2009**, *57*, 517–524. [CrossRef] [PubMed]

foods

MDPI

Article

Enzyme-Linked Aptamer Kits for Rapid, Visual, and Sensitive Determination of Lactoferrin in Dairy Products

Fan Zhang [1,2], Hongxia Du [1,2], Linsen Li [3], Tengfei Li [4], Jing Wang [5], Zilei Chen [1,2], Mengmeng Yan [1,*], Chao Zhu [1,2,*] and Feng Qu [3]

1 Institute of Quality Standard and Testing Technology for Agro-Products, Shandong Academy of Agricultural Sciences, Jinan 250100, China
2 Shandong Provincial Key Laboratory Test Technology on Food Quality and Safety, Jinan 250100, China
3 Key Laboratory of Molecular Medicine and Biotherapy, School of Life Science, Beijing Institute of Technology, Beijing 100081, China
4 School of Life Sciences and Food Engineering, Hebei University of Engineering, Handan 056038, China
5 School of Materials Science and Engineering, Hebei University of Engineering, Handan 056038, China
* Correspondence: ynky202@163.com (M.Y.); ndytzhuchao@126.com (C.Z.)

Abstract: Lactoferrin (Lf), as a popular nutritional fortification in dairy products, has the ability regulate the body's immune system and function as a broad-spectrum antibacterial, which is of great significance to the growth and development of infants and children. Herein, an indirect competitive enzyme-linked aptamer assay (ELAA) kit was established for rapid, sensitive, and visual determination of Lf in dairy products. In the construction, the Lf aptamer was conjugated with horseradish peroxidase (HRP) as the recognition probe and aptamer complementary strand (cDNA) were anchored onto the microplate as the capture probe. The recognition probes were first mixed with a sample solution and specifically bound with the contained Lf, then added into the microplate in which the free recognition probes in the mixture were captured by the capture probe. After washing, the remaining complex of cDNA/Aptamer/HRP in the microplate was conducted with a chromogenic reaction through HRP, efficiently catalyzing the substrate 3, 3′, 5, 5′-tetramethylbenzidine (TMB), therefore the color shade would directly reflect Lf concentration. Under the optimization conditions, a good linear relationship (R^2, 0.9901) was obtained in the wide range of 25–500 nM with the detection limit of 14.01 nM and a good specificity, as well as the reliable recoveries. Furthermore, the ELAA kits achieved the Lf determination with an accuracy of 79.71~116.99% in eleven samples, which consisted of three kinds of dairy products: including goat milk powder, cow milk powder, and nutrition drop. Moreover, the results were also validated by the high-performance capillary electrophoresis (HPCE) method. The ELAA kit provides a simple and convenient determination for Lf in dairy products, and it is highly expected to be commercialized.

Keywords: aptamer; lactoferrin determination; enzyme-linked aptamer assay; colorimetric; capillary electrophoresis

Citation: Zhang, F.; Du, H.; Li, L.; Li, T.; Wang, J.; Chen, Z.; Yan, M.; Zhu, C.; Qu, F. Enzyme-Linked Aptamer Kits for Rapid, Visual, and Sensitive Determination of Lactoferrin in Dairy Products. *Foods* **2022**, *11*, 3763. https://doi.org/10.3390/foods11233763

Academic Editor: Thierry Noguer

Received: 10 October 2022
Accepted: 18 November 2022
Published: 22 November 2022

Publisher's Note: MDPI stays neutral with regard to jurisdictional claims in published maps and institutional affiliations.

Copyright: © 2022 by the authors. Licensee MDPI, Basel, Switzerland. This article is an open access article distributed under the terms and conditions of the Creative Commons Attribution (CC BY) license (https://creativecommons.org/licenses/by/4.0/).

1. Introduction

Lactoferrin (Lf) is an iron-binding glycoprotein with a molecular weight of about 80 kDa, which is abundant in mammalian milk and plays an important role in both innate immunity and the growth of neonates, infants and young children [1,2]. Until now, Lf has been extensively studied and is demonstrated to have a wealth of functions such as broad-spectrum antibacterial [3], antiviral infection [4], regulating the body's iron balance [5], mediating the production of bone marrow cells [6], inhibiting human tumor cells [7], effectively treating diseases in synergy with various antibiotics and antifungal agents [8], etc. Due to its efficacy, Lf has been one of the fastest developing nutritional additives in recent years, and its products are increasingly used in the processing of dairy products, especially common in milk powder which is crucial for non-breastfed neonates and infants [9–11]. Hence, the sensitive detection for LF is of great significance in the fight against food adulteration and in support of market surveillance, and eventually for ensuring its function to be available in the organism. In China, the "Food Nutrition Fortification Use Standard (GB14880-2012)" stipulates that the maximum allowable use of Lf in infant formula is 1.0 g/kg, while few National Standards related to Lf determination in dairy products have been established.

The detection methods for Lf in dairy products are mainly classified as physicochemical techniques, biosensor assays, and immunoassays [12]. Among the physicochemical techniques, high-performance liquid chromatography (HPLC)-derived methods have been continuously and widely studied and applied, such as reverse-phase HPLC [13], size exclusion-HPLC [14], HPLC-fluorescence [15], and HPLC-tandem mass spectrometry (HPLC-MS/MS) [16]. Besides, electrophoresis methods, especially high-performance capillary electrophoresis (HPCE), are also attracting considerable attention with their advantages of high separation efficiency and fast analysis speed [17–19]. Although the physicochemical instrumental methods are generally the traditional methodology, some aspects involving inconvenient portability, the complicated pre-treatment, requirement of expensive instruments and professionals, sometimes negatively affect their application. Recently, biosensing technologies such as electrochemical sensors [20,21] and the surface plasmon resonance [22] method have been employed due to their simplicity, accuracy and sensitivity, and practicality. For example, Ebrahimi et al. [23] constructed an electrochemical sensor based on the Lf adsorption by the MOP self-assembled monolayer for Lf determination in colostrum samples with a low detection limit of about 65.2 nM. These methods still contained some shortcomings such as low throughput, poor reproducibility, and low environmental applicability, despite the biosensors being simple and having the sensitivity to determine Lf. The antibody-based immunoassays have been widely integrated with handy devices, such as enzyme-linked immunosorbent assay (ELISA) [24], microfluidic chip [25], and strips [26], but the process of antibody preparation requires animal immunization experiments and has a long development period. Given the diverse functions of Lf and the existing detection dilemma, it is of great significance to establish simple, convenient, and efficient methods for Lf determination.

Recently, Qu's group obtained an aptamer against Lf using CE-SELEX with an equilibrium dissociation constant (K_D) of 20.74 nM [27], which provided a basis for the establishment of an aptamer biosensor for Lf determination. An Aptamer is a short single-stranded nucleotide sequence that binds to a target molecule with high specificity and affinity, selected from a random oligonucleotide library by systematic evolution of ligands by exponential enrichment (SELEX) [28,29]. Known as "chemical antibodies", aptamer has the advantages of simple preparation, stable properties, a wide range of targets, low immunogenicity, and easy modification compared with the antibody [30,31]. Currently, with the advancement of aptamer screening and synthesis methods, its downstream applications are becoming more and more extensive: such as medicine, life science, and bioanalytical science [32,33]. Among the numerous aptasensors, the horseradish peroxidase (HRP) enzyme-linked aptamer assay (ELAA) kit [34,35], similar to the well-established and approved antibody-based ELISA kit in the market, is one of the most anticipated applications because of its simplicity, rapidity, and visualization.

In this study, an indirect competitive ELAA kit was established for Lf determination in dairy products. The HRP-labeled aptamers were reacted sequentially with the measured samples and the aptamer's complementary strand (cDNA) immobilized on the microplate to rapidly determine Lf content through the color change based on the enzyme-catalyzed substrate. This sequential reaction avoids excessive binding of the aptamers to cDNA, thus ensuring full-coverage recognition of Lf. The ELAA kit presented high sensitivity and stability in Lf determination, as well as good spiked recoveries, and was successfully applied to the determination of Lf in thirteen samples obtained from the market. In addition, the workflow of the ELAA kit could be accomplished through the convenient "mix-wash-detect" operation for an entire time within 60 min, which exhibited a certain potential as the rapid and efficient method for Lf determination in dairy products.

2. Experimental Sections

2.1. Reagents, Materials and Instruments

Lf standards (from bovine milk), bovine serum albumin (BSA), α-lactalbumin (α-la), β-lactoglobulin (β-lg), transferrin (Tf) were purchased from Sigma-Aldrich LLC (Beijing, China). Streptavidin (SA) and biotin-modified HRP were purchased from Aladdin Reagent (Shanghai, China). The 96-well plate was purchased from Thermo Fisher Scientific (Waltham MA, USA); single component 3, 3′, 5, 5′-tetramethylbenzidine (TMB), TE buffer, PBS buffer (pH 7.2) were purchased from Solarbio Biological Technology Company (Beijing, China). H_2SO_4 and Tween-20 were supported by Beijing Reagent Plant (Beijing, China). Polyethylene glycol dodecylether (Brij 35) was purchased from Macklin Inc. (Beijing, China). The Aptamer sequence of biotin-TGGTGCTGCCCCCCTAGTCTCCGGCTGCTTCTTGG and its complementary chain sequence of biotin-CCAAGAAGCAGCCGGAGACTAGGGG CAGCACCA were ordered and purified by Shanghai Sangon Biotechnology Co., Ltd. (Shanghai, China). The capillary was obtained by Sino Sumtech (Handan, Hebei, China).

The HPCE model is WooKing HPCE512 produced by Hanon Group (Jinan, China) and equipped with a UV detector (214 nm). The absorbance at 450 nm was measured by a Microplate reader (Multiskan™ FC, Thermo Fisher Scientific, Waltham MA, USA). The samples were prepared using the circular oscillator (MS 3 digital, IKA, Staufen, Germany) and high-speed refrigerated centrifuge (Neofuge 15R, Heal Force, ShangHai, China). The ultrapure water used throughout the experiments was purified by a Milli-Q system (Bedford MA, USA) and had a resistivity of 18.2 MΩ cm. Before use, all solutions were filtered through a 0.22 μm filter (Boston, MA, USA).

2.2. The Preparation of Capture Probe

First, SA was diluted in carbonate buffer (50 mM, pH 9.6) and 50 μL of SA solution was added into each well of the 96-well microplate and incubated for 1 h at room temperature which was controlled at about 25 °C by air conditioning. After that, the wells were poured and washed three times with PBS-T buffer (1×PBS buffer with 0.05% Tween-20) and patted dry. To reduce non-specific adsorption, the 96-well plate was blocked by adding 300 μL of BSA solution to the wells that were not completely coated with SA, and the wells were blocked for 1 h. Then, the wells were washed three times with PBS-T, each time retaining the washing solution for 1 min. As a result, the SA-modified microplate was obtained.

The biotin-modified cDNA chain was diluted with TE buffer (pH 8.0), and 50 μL solution was added to each well of the SA-modified microplate and incubated for 10 min. After that, the solution in the wells was poured out and washed three times with PBS-T buffer and patted dry. The prepared capture probe was eventually stored in a refrigerator at 4 °C for later use.

2.3. The Preparation of Recognition Probe

The SA-modified HRP (1 mg/mL) was diluted to the desired concentration using PBS (pH 7.2), and then was mixed with biotin-labeled aptamer at a ratio of 7:3. The mixture was added into a centrifuge tube and incubated for 30 min on a rocking bed. In the end, the prepared recognition probe was stored in a refrigerator at 4 °C for later use.

2.4. The Operation of ELAA Kit

The recognition probes were first mixed with the Lf samples in equal volumes for about 5 min at room temperature. Then, 100 μL of the mixture solution was loaded into each well of the microplate to incubate with a capture probe for about 30 min. The microplate was washed three times with PBS-T and patted dry. In subsequent steps, 100 μL of TMB solution was separately added into each well, and after 5 min 50 μL of 2 M H_2SO_4 solution was added to terminate the enzyme catalysis. The absorbances (A_{450}) were scanned by a

Premium reader, the relative A_{450} was used to calculate the Lf content, which was obtained by the formula of $(A_b - A_s)/A_b$ where A_b represented the A_{450} value of blank and A_s represented the A_{450} value of the sample.

2.5. Evaluation of Selectivity and Stability

In the selectivity assay, the concentration of Lf protein was 300 nM, and the concentration of all other proteins was 3 µM. The stability assay was verified within 15 days after keeping the constant of capture probes and recognition probes. During the experiments, the temperature of 25 °C and the Lf concentration of 100 nM were kept constant. The assays were performed under optimal experimental conditions, and the final absorbance value at 450 nm was measured to evaluate the method specificity.

2.6. Treatment of Dairy Products

First, 0.05 g of powder or 0.05 mL of nutrition drops was dissolved in 1 mL of acetic acid solution (50 mM) and mixed well. The mixture was centrifuged at 8000 r/min for 10 min. Three layers of samples were obtained after centrifugation, from top to bottom, fat layer, clear liquid layer, and precipitation layer. The 200 µL of the clear liquid layer was aspirated with a syringe and subsequently filtered through a 0.22 µm cellulose acetate membrane to obtain the sample solution. The sample solution was analyzed with an ELAA kit.

2.7. HPCE Assay

The HPCE for Lf determination was referenced by Li et al. [18]. The bare capillary (50 µm i.d.) with a total length of 32.6 cm (effective length 20.3 cm) was rinsed with 1 M NaOH for 25 min, and water for 5 min at the first use. In the HPCE assay, the sequential rinses for capillary were required between each run for 3 min with 1 M NaOH, water, and running buffer solution, in which the composition was 40 mM NaH_2PO_4, 40 mM H_3PO_4, and 5 mM Brij 35. The sample was injected at 0.5 psi for 20 s or the required time. The detection wavelength was set to 214 nm. During the separation, the high voltage of 12 kV (the inlet as the anodes) and the temperature of 22 °C were maintained.

3. Results and Discussion

3.1. Working Principle of the ELAA Kit

The working principle of the ELAA kit was illustrated in Figure 1 as follows. Initially, the capture probe of the biotin-labeled cDNA was anchored onto the surface of both streptavidin (SA)-coated and BSA pre-blocked polystyrene microplate wells through the specific interaction between SA and biotin. The recognition probe was constructed through the conjugation between SA-modified HRP and biotin-labeled aptamer. In the absence of Lf in a sample, there is no binding event between the aptamer and Lf, therefore the recognition probe would subsequently hybridize with the capture probe through the base complementary pairing between the aptamer and cDNA. After washing, the rich complexes of aptamer/biotin-SA/HRP remained in the wells, in which numerous HRPs were available for the added reporters of TMB to be adequately catalyzed, yielding a sharp color change from pale to dark yellow. In the presence of Lf in the sample, the aptamer elements-affiliated recognition probe would first target Lf analytes due to their specific recognition and high-affinity binding, which thereby lead to a decrease in the number of unbound recognition probes. Correspondingly, fewer complexes of aptamer/biotin-SA/HRP remained and caused insufficient catalysis of the TMB along with the color change at different degrees, in which color shades were negatively correlated with Lf concentrations in the sample. The displayed ELAA kit was performed by employing a convenient "mix-wash-detect" workflow and a sequential reaction-based indirect competition that ensured full-coverage binding of Lf to achieve a robust determination, and reduced the feasibility of the accessibility to nonspecific binding sites on the surface. The results of the yellow shade were directly observed by the naked eye and the absorbance (A_{450}) was scanned by a Premium reader, which enabled achieving semi- and quantitative determination, respectively.

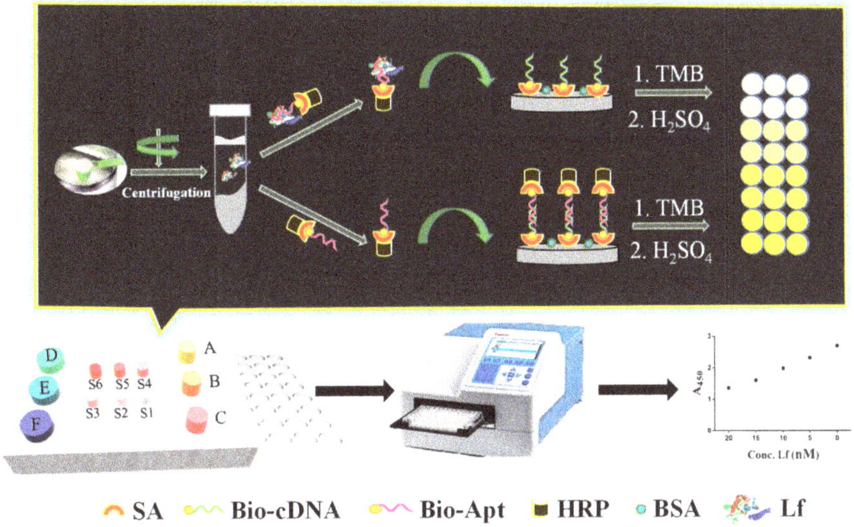

Figure 1. The construction and schematic of the ELAA kit for Lf determination.

3.2. Optimization of Experimental Conditions

The careful optimizations of a series of parameters were conducted. These optimized conditions ensured that the final color was relatively easy to distinguish and that the rapid and accurate determination of the Lf was fully guaranteed. First, three aspects in the constructions of the capture probe were optimized involving the concentrations of SA modified onto the bottoms of a microplate, the concentrations of BSA blocking solution, and the concentrations of the cDNA. Figure 2A displayed that the SA concentration reached 10 µg/mL, and the A_{450} value no longer increased with the increased concentration, but leveled off, which the concentration then ensured that the actual SA amount coated onto the microplate reached saturation, and was thus selected. Figure 2B showed that the A_{450} value of 0.5% BSA was fundamentally the same as that of unblocked, indicating that the microplate was coated successfully. In addition, the A_{450} value decreased as the BSA concentration increased, indicating that the high concentration of BSA would instead reduce the kit's sensitivity. The 0.5% BSA was still selected to prevent non-specific adsorption despite its seeming inability. Similarly, the cDNA concentration of 50 nM (Figure 2C) presented a relatively high and stable A_{450} value and was selected. Particularly, the high cDNA concentration resulted in a slightly reduced A_{450} value, most likely due to the mismatched binding between cDNAs.

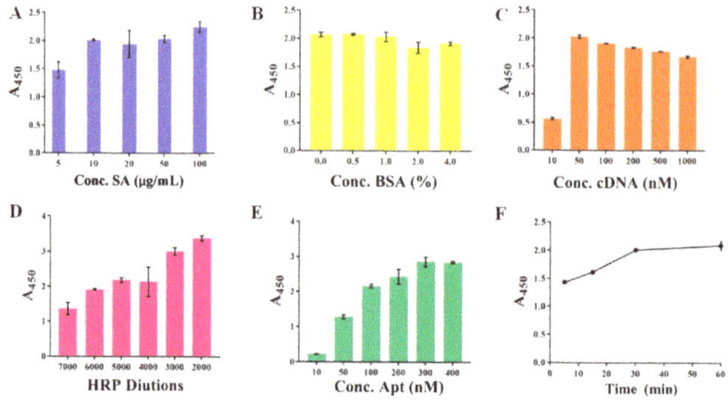

Figure 2. The optimizations of capture probe and recognition probe involving the concentrations (**A**) of SA, the concentrations (**B**) of BSA blocking solution, the concentrations (**C**) of the cDNA, the concentrations of SA-HRP (**D**) and biotin-Apt (**E**), as well as their incubation time (**F**).

Subsequently, the constructions of the recognition probe were optimized, including the concentrations of SA-HRP and biotin-Apt, as well as their incubation time, which determined the amount of the bound Lf and the catalyzed TMB, thereby significantly affecting the kits analytical performance. Figure 2D showed that the A_{450} value increased with the increasing SA-HRP concentration, in which the high concentration at the dilution of 1:2000 and 1:3000 gave excessively high A_{450} values (>3.0) that reached the higher limit of the detector, whereas the dilution of 1:5000 presented a relatively suitable A_{450} value of 2.5 and its color was easy to distinguish with the naked eye, thereby employed in the following experiments. As anticipated, more biotin-Apt probes could provide more opportunities for accessibility of the cDNAs, resulting in the higher signal output of the A_{450} value (Figure 2E), and the concentration from the moderate part of 200 nM was selected. Figure 2F displayed the incubation time between SA-HRP and biotin-Apt mounted up to 30 min, the absorbance approached a relatively higher value, and in the next 30 min the increase was only 5.4%. Thus, we chose 30 min as the optimal incubation time.

Furthermore, the procedures in the detection operation were optimized, including the incubation buffer, the binding time of the recognition probes separately to the Lf target and to subsequent capture probes, the added volume of TMB, and the termination time. The results displayed that: the PBS buffer presented sufficient signal output whereas the other three buffers did not (Figure 3A); A_{450} increased obviously with the increasing two probes' binding time from 0–30 min, whereas in the next 30 min the increase was only 5.4% (Figure 3B); the absorbance at the binding time of 5 min decreased to a relatively low value (Figure 3C) and then became plateau due to the high affinity between the aptamer and Lf protein; the TMB volume of 100 μL (Figure 3D) and the termination time of 5 min guarantee the relatively suitable absorbance (Figure 3E), whereas although a longer termination time could yield larger A_{450} values, the corresponding excessively dark color was not easily distinguished by the naked eye; thereby, these parameters were employed in the following experiments.

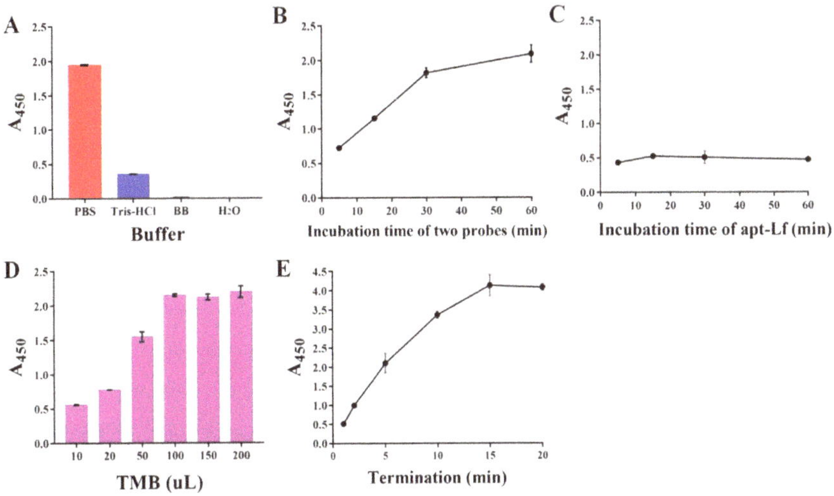

Figure 3. The optimizations of the procedures in detection operation involving the incubation buffer (**A**), the binding time of the recognition probes separately to subsequent capture probes (**B**) and to the Lf target (**C**), the added volume (**D**) of TMB, and the termination time (**E**).

3.3. *Performances of ELAA Kit*

Under the optimization conditions, the sensitivity was evaluated using a standard Lf solution of different concentrations ranging from 25 to 500 nM. As shown in Figure 4A (insert picture), with the concentration of Lf increased, the blue color of the solution gradually becomes evidently lighter with the naked eye. However, the direct competition-based assay displayed no significant difference in color. A good calibration curve was fitted with a reliable correlation coefficient (R^2) of 0.9901 based on the linear relationship between the relative A_{450} values (vertical coordinate) and the Lf concentration (horizontal coordinate). Error bars were obtained based on three parallel measurements at different concentrations. Meanwhile, the regression equation was obtained by $y = 0.564x - 0.7074$, along with a limit of detection (LOD) of 14.01 nM (1.12 mg/L) based on the 3σ ($\sigma = \alpha/k$, $\alpha = 2.6345$ is the standard deviation of the blank signal, $k = 0.564$ is the slope of the standard curve), whereas the direct competition-based ELAA showed a poor R^2 of 0.7193 and a higher LOD of 347.14 nM. Therefore, it could be concluded that the indirect competition-based ELAA kit was capable of determining Lf quantitatively. Although the LOD was at an intermediate level compared to other reported ELISA methods (Table S1) [26,36–38], it fully meets China's GB14880-2012 which shall not exceed 1.0 g/L, along with a convenience "mix-wash-detect" operation. In addition, these ELISA methods are based on the immunoreaction between Lf and its antibody, which are expensive reagents and complicated to prepare.

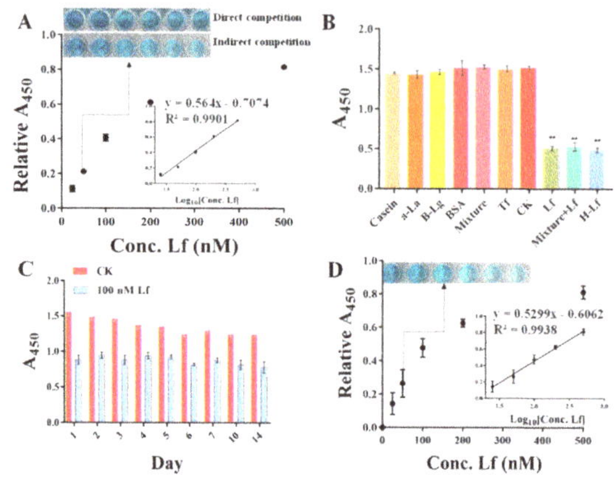

Figure 4. The performances of ELAA kit involve sensitivity (**A**), specificity (** Represents a significant difference, **B**), and stability (**C**), and the calibration curve and sensitivity in spiked matrix (**D**).

To verify the selectivity of the ELAA kit and prevent the occurrence of false positives, other proteins were evaluated simultaneously by the ELAA kit, including casein accounting for a large proportion in milk, and the main whey proteins of bovine serum albumin (BSA), α-lactalbumin (α-la), β-lactoglobulin (β-lg), and their mixture, as well as the blank samples as controls (CK). As shown in Figure 4B, the wells of only Lf and Lf-contained a mixture which presented similar A_{450} nm values, whereas the blank sample and other individual proteins, as well as their mixture and even with 10-fold concentration, provided higher and similar absorbance, indicating that the ELAA kit had a good selectivity. In addition, the specificity was also evaluated using other proteins of human lactoferrin (H-Lf) that belong to different genera, and Transferrin (Tf) that also belongs to the transferrin family. Figure 4B depicts that the ELAA kit presented a similar recognition to H-Lf, perhaps due to their similar structures, as well as a high specificity in transferrin recognition. These results all indicated that the ELAA can be used in the specific recognition and determination of Lf.

To assess the reproducibility of the ELAA kit, we performed three groups of experiments including the intra-batch assay, intra-day assay, and inter-day assay. The A_{450} values were measured using three Lf standards of 60, 80, and 100 nM. The coefficient of variation (CV) is a relative measure of dispersion and is often used as an important indicator for reproducibility. A higher CV value means a more significant detection error and a lower value means a more stable result. As listed in Table 1, the calculated average CVs of intra-day, inter-day, and intra-batch assays for the ELAA kit were respectively 0.01–0.07, demonstrating the reproducibility of the ELAA kit based on the low-value. Moreover, the stability of the ELAA kit was verified within 14 days by utilizing the same lot of kits. Figure 4C showed that A_{450} values were not significantly changed with a good relative standard deviation (RSD) of 5.72%, indicating that the ELAA kit has good stability and can be used to prepare kits for long-term use for a certain period of time.

Table 1. Reproducibility of the ELAA kit.

Conc. Lf (nM)	Intra-Day Assay			Inter-Day Assay			Intra-Batch Assay		
	Mean [a]	SD [b]	CV [c]	Mean	SD	CV	Mean	SD	CV
60	1.34	0.09	0.07	1.37	0.01	0.01	1.11	0.04	0.03
80	1.04	0.01	0.01	1.16	0.04	0.03	0.92	0.03	0.03
100	0.80	0.05	0.06	0.82	0.01	0.02	0.67	0.01	0.02

[a] Values represent the average of the A_{450} values of the parallel samples (n = 3). [b] Values represent the standard deviation of the parallel results (n = 3). [c] CV = SD/mean.

3.4. Calibration Curve and Sensitivity in Spiked Matrix

To assess the applicability and accuracy of the ELAA kit, the calibration curve and sensitivity in the milk powder matrix were conducted. We chose one commercially available milk powder labeled with no Lf as the blank sample that was also confirmed by using the HPCE method (Figure 5 insert picture). The milk powder was subjected to a simple pre-treatment. After centrifugation, the intermediate layer of liquid was removed to be used as a matrix solution, by which the stocked solution of Lf standard was diluted to different concentrations ranging from 25 to 500 nM. Though assessed by the ELAA kit, Figure 4D showed that the relative A_{450} values in the microwells gradually became larger as the Lf concentration increased, and the calibration curve was obtained as y = 0.5299x − 0.6062 with the LOD of 17.08 nM (1.36 mg/kg). Furthermore, the R^2 in matrixes was greater than 0.99, showing good linearity in the analytical range. Such high sensitivity could determine potential Lf content in milk powders. However, a slightly different slope between regression equations obtained by "sample matrix" and "standard solutions" indicates a matrix effect that can affect the results. Thus, the recovery experiment was also conducted using the matrix calibration curve by spiking the Lf standard into milk powder before pre-treatment to five distinct concentrations. Table 2 showed that the recovery rates of samples were varied in the range of 95–107%, as well as their RSD values of less than 2%. The results confirmed the reliability and accuracy of the ELAA kit to determine Lf sensitively in the milk powder matrix.

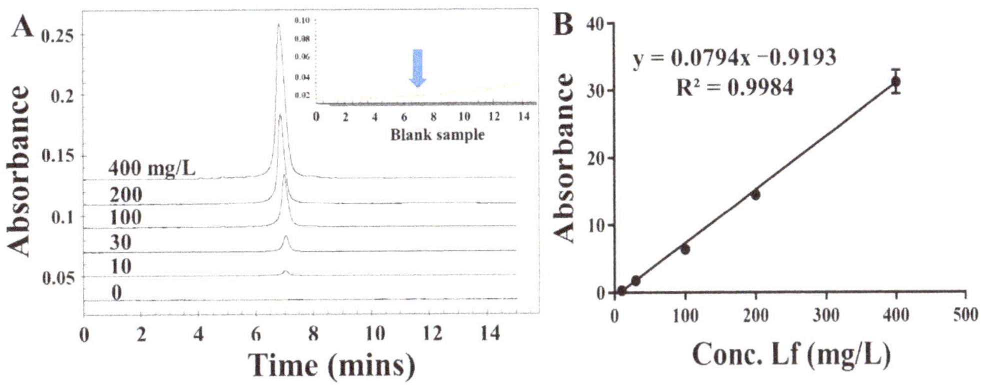

Figure 5. The electropherograms of different concentrations of LF standard (**A**) and the calibration curve (**B**).

Table 2. Recovery tests in milk powder matrix using the ELAA kit.

Spiked (nM)	Found (nM)	Recovery (%)	RSD (%)
50	53.40	106.80	1.94
100	101.90	101.90	1.41
150	142.70	95.13	1.98

3.5. Application Evaluation in Real Samples

To evaluate the actual application of the ELAA kit, we determined the Lf content in thirteen samples that were purchased from the market and consisted of three kinds of dairy products, including goat milk powder (G1-G6), cow milk powder (C1-C6), nutrition drop (N-1), as well as with different Lf concentrations. Furthermore, these samples were also subjected to the verified assay by employing the HPCE method (Figure 5), and the CE results were regarded as the yardstick. Compared with the HPCE results (Table 3), the eleven results in the ELAA kit showed that the Lf concentrations are fundamentally consistent with HPCE values, along with the accuracy rates of these samples limited in the range of 79.71%~116.99%, as well as their RSD values of less than 3.61%. In particular, the blank samples are all tested accurately, indicating a good performance of the ELAA kit in real samples. There are two samples with slightly large differences in test results (138.34% and 146.43%), which may be due to their more complex matrix composition. However, the ELAA kit took only 60 min to accomplish analysis for all samples by only performing a single operation, which was much shorter than that required by the HPCE of about 12 h (including washing time) for all samples.

Table 3. Results of thirteen samples using the HPCE method and ELAA kit.

Sample	Label (nM)	HPCE (nM)	ELAA (nM)	Accuracy (%)	RSD (%)
C-1	0	0	0	100	0.03
C-2	0	0	0	100	0.39
C-3	420	351.5	403.1	114.70	1.41
C-4	480	521.1	499.1	95.78	2.26
C-5	600	594.6	599.3	100.78	0.73
C-6	360	220.2	304.6	138.34	2.81
G-1	0	0	0	100	0.43
G-2	0	0	0	100	0.23
G-3	540	572.1	469.5	82.07	2.82
G-4	510	402.5	381.6	94.81	0.13
G-5	78	122.3	179.1	146.43	0.05
G-6	600	288.7	337.7	116.99	0.03
N-1	9000	8961.8	7144.0	79.71	3.61

4. Conclusions

In this study, an enzyme-linked aptamer colorimetry kit was constructed to determine Lf in dairy products. To achieve a robust determination, the sequential reaction was employed between the recognition probes with the measured samples and subsequent capture probes, in which the indirect competition allowed full-coverage binding of the aptamers to Lf and avoided excessive binding. By a "mix-wash-detect" operation, the proposed ELAA kit demonstrates low limits of detection, good stability, and specificity under the optimized conditions, as well as a good calibration curve and high recovery rates in the spiked matrix. Furthermore, it also achieved Lf determination in eleven samples for three kinds of dairy products with different Lf concentrations, and the results were confirmed by the CE method with high accuracy. In the absence of rapid and accurate analysis for Lf determination in dairy products, our aptamer-based ELAA kit offers considerable advantages in terms of convenience, detection speed and time (~60 min), throughput (96-wells), and cost of analysis (~36 USD).

Supplementary Materials: The following supporting information can be downloaded at: https://www.mdpi.com/article/10.3390/foods11233763/s1, Table S1: The comparison of ELISA and ELAA.

Author Contributions: F.Z.: Investigation, Validation, Data curation, Formal analysis. H.D.: Data curation, Validation, Formal analysis. L.L.: Data curation, Formal analysis. T.L.: Formal analysis. J.W.: Formal analysis. Z.C.: Writing-review & editing. M.Y.: Writing original draft, Data curation, Funding acquisition, Project administration. C.Z.: Writing original draft, Data curation, Funding acquisition, Project administration. F.Q.: Writing-review & editing. All authors have read and agreed to the published version of the manuscript.

Funding: We thank the support of the Natural Science Foundation of Shandong Province (no. ZR2022QB207), National Natural Science Foundation of China [(no. 32102088), Supported by the Taishan Scholars Program, the Agricultural Scientific and Technological Innovation Project of Shandong Academy of Agricultural Sciences (CXGC2021B14, CXGC2022E05), and the Young Elite Scientist Sponsorship Program of Beijing Association for Science and Technology.

Data Availability Statement: All related data and methods are presented in this paper. Additional inquiries should be addressed to the corresponding author.

Conflicts of Interest: The authors declare no conflict of interest.

References

1. Adlerova, L.; Bartoskova, A.; Faldyna, M. Lactoferrin: A review. *Vet. Med.* **2008**, *53*, 457–468. [CrossRef]
2. Dall'Agnola, A.; Tomé, D.; Kaufman, D.A.; Tavella, E.; Pieretto, M.; Messina, A.; De Luca, D.; Bellaiche, M.; Mosca, A.; Piloquet, H.; et al. Role of Lactoferrin in Neonates and Infants: An Update. *Am. J. Perinatol.* **2018**, *35*, 561–565. [CrossRef] [PubMed]
3. Sienkiewicz, M.; Jaśkiewicz, A.; Tarasiuk, A.; Fichna, J. Lactoferrin: An overview of its main functions, immunomodulatory and antimicrobial role, and clinical significance. *Crit. Rev. Food Sci. Nutr.* **2021**, *62*, 6016–6033. [CrossRef] [PubMed]
4. Salaris, C.; Scarpa, M.; Elli, M.; Bertolini, A.; Guglielmetti, S.; Pregliasco, F.; Blandizzi, C.; Brun, P.; Castagliuolo, I. Protective Effects of Lactoferrin against SARS-CoV-2 Infection In Vitro. *Nutrients* **2021**, *13*, 328. [CrossRef]
5. Mancinelli, R.; Rosa, L.; Cutone, A.; Lepanto, M.S.; Franchitto, A.; Onori, P.; Gaudio, E.; Valenti, P. Viral Hepatitis and Iron Dysregulation: Molecular Pathways and the Role of Lactoferrin. *Molecules* **2020**, *25*, 1997. [CrossRef]
6. Naot, D.; Grey, A.; Reid, I.R.; Cornish, J. Lactoferrin-A Novel Bone Growth Factor. *Clin. Med. Res.* **2005**, *3*, 93–101. [CrossRef]
7. Ramírez-Rico, G.; Drago-Serrano, M.E.; León-Sicairos, N.; de la Garza, M. Lactoferrin: A Nutraceutical with Activity against Colorectal Cancer. *Front. Pharmacol.* **2022**, *13*, 855852. [CrossRef]
8. Li, Y.-Q.; Guo, C. A Review on Lactoferrin and Central Nervous System Diseases. *Cells* **2021**, *10*, 1810. [CrossRef]
9. Bobreneva, I.V.; Rokhlova, M.V. Lactoferrin: Properties and application—A review. *Theory Pract. Meat Process.* **2021**, *6*, 128–134. [CrossRef]
10. Wang, B.; Timilsena, Y.P.; Blanch, E.; Adhikari, B. Lactoferrin: Structure, function, denaturation and digestion. *Crit. Rev. Food Sci. Nutr.* **2019**, *59*, 580–596. [CrossRef]
11. Niaz, B.; Saeed, F.; Ahmed, A.; Imran, M.; Maan, A.A.; Khan, M.K.I.; Tufail, T.; Anjum, F.M.; Hussain, S.; Suleria, H.A.R. Lactoferrin (LF): A natural antimicrobial protein. *Int. J. Food Prop.* **2019**, *22*, 1626–1641. [CrossRef]
12. Zhang, Y.; Lu, C.; Zhang, J. Lactoferrin and Its Detection Methods: A Review. *Nutrients* **2021**, *13*, 2492. [CrossRef] [PubMed]
13. Tsakali, E.; Chatzilazarou, A.; Houhoula, D.; Koulouris, S.; Tsaknis, J.; Van Impe, J. A rapid HPLC method for the determination of lactoferrin in milk of various species. *J. Dairy Res.* **2019**, *86*, 238–241. [CrossRef]
14. Mohamed, H.G.; Al-Ghobashy, M.A.; Fouad, M.A.; Zaazaa, H.S. Quality Assessment of Lactoferrin in some Marketed Nutraceuticals Derived from Milk using Validated Analytical Methods. *ChemistrySelect* **2020**, *5*, 14816–14825. [CrossRef]
15. Pang, J.; Xiao, Q.; Yan, H.; Cao, Y.; Miao, J.; Wang, S.; Li, X.; Li, H.; Cheng, Z. Bovine Lactoferrin Quantification in Dairy Products by a Simple Immunoaffinity Magnetic Purification Method Coupled with High-Performance Liquid Chromatography with Fluorescence Detection. *J. Agric. Food Chem.* **2019**, *68*, 892–898. [CrossRef] [PubMed]
16. Jia, Y.; Lu, Y.; Wang, X.; Yang, Y.; Zou, M.; Liu, J.; Jin, W.; Wang, X.; Pang, G.; Huang, L.; et al. Mass spectrometry based quantitative and qualitative analyses reveal N-glycan changes of bovine lactoferrin at different stages of lactation. *LWT* **2021**, *147*, 111626. [CrossRef]
17. Chen, H.; Wang, Z.; Fan, F.; Shi, P.; Xu, X.; Du, M.; Wang, C. Analysis Method of Lactoferrin Based on Uncoated Capillary Electrophoresis. *eFood* **2021**, *2*, 147–153. [CrossRef]
18. Li, J.; Ding, X.; Chen, Y.; Song, B.; Zhao, S.; Wang, Z. Determination of bovine lactoferrin in infant formula by capillary electrophoresis with ultraviolet detection. *J. Chromatogr. A* **2012**, *1244*, 178–183. [CrossRef]
19. Mao, K.; Du, H.; Bai, L.; Zhang, Y.; Zhu, H.; Wang, Y. Poly (2-methyl-2-oxazoline) coating by thermally induced immobilization for determination of bovine lactoferrin in infant formula with capillary electrophoresis. *Talanta* **2017**, *168*, 230–239. [CrossRef]
20. Ebrahimi, F.; Mozaffari, S.A.; Amoli, H.S. New method for evaluating the concentration of lactoferrin protein in infant formula using electrochemical sensors technology. *Innov. Food. Technol.* **2020**, *7*, 627–640. [CrossRef]
21. Lu, Y.; Ke, H.; Wang, Y.; Zhang, Y.; Li, H.; Huang, C.; Jia, N. A ratiometric electrochemiluminescence resonance energy transfer platform based on novel dye BODIPY derivatives for sensitive detection of lactoferrin. *Biosens. Bioelectron.* **2020**, *170*, 112664. [CrossRef] [PubMed]
22. Tomassetti, M.; Martini, E.; Campanella, L.; Favero, G.; Sanzò, G.; Mazzei, F. Lactoferrin determination using flow or batch immunosensor surface plasmon resonance: Comparison with amperometric and screen-printed immunosensor methods. *Sensors Actuators B Chem.* **2013**, *179*, 215–225. [CrossRef]
23. Ebrahimi, F.; Amoli, H.S.; Mozaffari, S.A. Impedimetric and single-frequency capacitance spectroscopy strategy in label-free rapid screening of lactoferrin. *Sens. Actuators B Chem.* **2021**, *354*, 131107. [CrossRef]
24. Ostertag, F.; Sommer, D.; Berensmeier, S.; Hinrichs, J. Development and validation of an enzyme-linked immunosorbent assay for the determination of bovine lactoferrin in various milk products. *Int. Dairy J.* **2021**, *125*, 105246. [CrossRef]
25. Kudo, H.; Maejima, K.; Hiruta, Y.; Citterio, D. Microfluidic Paper-Based Analytical Devices for Colorimetric Detection of Lactoferrin. *SLAS Technol. Transl. Life Sci. Innov.* **2019**, *25*, 47–57. [CrossRef]
26. Wang, R.; Wang, J.; Liu, H.; Gao, Y.; Zhao, Q.; Ling, S.; Wang, S. Sensitive immunoassays based on specific monoclonal IgG for determination of bovine lactoferrin in cow milk samples. *Food Chem.* **2020**, *338*, 127820. [CrossRef]
27. Zhu, C.; Li, L.; Yang, G.; Irfan, M.; Wang, Z.; Fang, S.; Qu, F. High-efficiency selection of aptamers for bovine lactoferrin by capillary electrophoresis and its aptasensor application in milk powder. *Talanta* **2019**, *205*, 120088. [CrossRef]
28. Ellington, A.D.; Szostak, J.W. In vitro selection of RNA molecules that bind specific ligands. *Nature* **1990**, *346*, 818–822. [CrossRef]

29. Tuerk, C.; Gold, L. Systematic Evolution of Ligands by Exponential Enrichment: RNA Ligands to Bacteriophage T4 DNA Polymerase. *Science* **1990**, *249*, 505–510. [CrossRef]
30. Röthlisberger, P.; Hollenstein, M. Aptamer chemistry. *Adv. Drug Deliv. Rev.* **2018**, *134*, 3–21. [CrossRef]
31. Wang, T.; Chen, C.; Larcher, L.M.; Barrero, R.A.; Veedu, R.N. Three decades of nucleic acid aptamer technologies: Lessons learned, progress and opportunities on aptamer development. *Biotechnol. Adv.* **2018**, *37*, 28–50. [CrossRef] [PubMed]
32. Wu, L.; Wang, Y.; Xu, X.; Liu, Y.; Lin, B.; Zhang, M.; Zhang, J.; Wan, S.; Yang, C.; Tan, W. Aptamer-Based Detection of Circulating Targets for Precision Medicine. *Chem. Rev.* **2021**, *121*, 12035–12105. [CrossRef]
33. Moutsiopoulou, A.; Broyles, D.; Dikici, E.; Daunert, S.; Deo, S.K. Molecular Aptamer Beacons and Their Applications in Sensing, Imaging, and Diagnostics. *Small* **2019**, *15*, e1902248. [CrossRef] [PubMed]
34. Svobodova, M.; Skouridou, V.; Jauset-Rubio, M.; Viéitez, I.; Fernández-Villar, A.; Alvargonzalez, J.J.C.; Poveda, E.; Bofill, C.B.; Sans, T.; Bashammakh, A.; et al. Aptamer Sandwich Assay for the Detection of SARS-CoV-2 Spike Protein Antigen. *ACS Omega* **2021**, *6*, 35657–35666. [CrossRef] [PubMed]
35. Zhang, D.-W.; Sun, C.-J.; Zhang, F.-T.; Xu, L.; Zhou, Y.-L.; Zhang, X.-X. An electrochemical aptasensor based on enzyme linked aptamer assay. *Biosens. Bioelectron.* **2012**, *31*, 363–368. [CrossRef]
36. Otnaess, A.-B.K.; Meberg, A.; Sande, H.A. Plasma Lactoferrin Measured by an Enzyme-Linked Immunosorbent Assay (ELISA). *Scand. J. Haematol.* **2009**, *31*, 235–240. [CrossRef] [PubMed]
37. Liu, L.; Kong, D.; Xing, C.; Zhang, X.; Kuang, H.; Xu, C. Sandwich immunoassay for lactoferrin detection in milk powder. *Anal. Methods* **2014**, *6*, 4742–4745. [CrossRef]
38. Dupont, D.; Arnould, C.; Rolet-Repecaud, O.; Duboz, G.; Faurie, F.; Martin, B.; Beuvier, E. Determination of bovine lactoferrin concentrations in cheese with specific monoclonal antibodies. *Int. Dairy J.* **2006**, *16*, 1081–1087. [CrossRef]

foods

MDPI

Article

Comparison of Chemiluminescence Enzyme Immunoassay (Cl-ELISA) with Colorimetric Enzyme Immunoassay (Co-ELISA) for Imidacloprid Detection in Vegetables

Rongqi Zhai [1,†], Ge Chen [1,†], Guangyang Liu [1], Xiaodong Huang [1], Xiaomin Xu [1], Lingyun Li [1], Yanguo Zhang [1], Donghui Xu [1,*] and A. M. Abd El-Aty [2,3]

1 Key Laboratory of Vegetables Quality and Safety Control, Laboratory of Quality & Safety Risk Assessment for Vegetable Products, Ministry of Agriculture and Rural Affairs, Institute of Vegetables and Flowers, Chinese Academy of Agricultural Sciences, Beijing 100081, China
2 Department of Pharmacology, Faculty of Veterinary Medicine, Cairo University, Giza 12211, Egypt
3 Department of Medical Pharmacology, Faculty of Medicine, Atatürk University, 25240 Erzurum, Turkey
* Correspondence: xudonghui@caas.cn; Tel.: +86-10-8210-6963
† These authors contributed equally to this work.

Abstract: Imidacloprid is one of the most commonly used insecticides for managing pests, thus, improving the quality and yield of vegetables. The abuse/misuse of imidacloprid contaminates the environment and threatens human health. To reduce the risk, a colorimetric enzyme-linked immunoassay assay (Co-ELISA) and chemiluminescence enzyme-linked immunoassay assay (Cl-ELISA) were established to detect imidacloprid residues in vegetables. The linear range of Co-ELISA ranged between 1.56 μg/L and 200 μg/L with a limit of detection (LOD) of 1.56 μg/L. The values for Cl-ELISA were 0.19 μg/L to 25 μg/L with an LOD of 0.19 μg/L, which are lower than those of Co-ELISA. Fortifying Chinese cabbage, cucumber, and zucchini with imidacloprid at 10, 50, and 100 μg/L yielded recoveries between 81.7 and 117.6% for Co-ELISA and at 5, 10, and 20 μg/L yielded recoveries range from 69.7 to 120.6% for Cl-ELISA. These results indicate that Cl-ELISA has a high sensitivity and a rapid detection time, saving cost (antigen and antibody concentrations) and serving as a more efficient model for the rapid detection of imidacloprid residue.

Keywords: imidacloprid; colorimetric assay; chemiluminescent assay; enzyme-linked immunoassay

Citation: Zhai, R.; Chen, G.; Liu, G.; Huang, X.; Xu, X.; Li, L.; Zhang, Y.; Xu, D.; Abd El-Aty, A.M. Comparison of Chemiluminescence Enzyme Immunoassay (Cl-ELISA) with Colorimetric Enzyme Immunoassay (Co-ELISA) for Imidacloprid Detection in Vegetables. *Foods* **2023**, *12*, 196. https://doi.org/10.3390/foods12010196

Academic Editor: Thierry Noguer

Received: 24 November 2022
Revised: 22 December 2022
Accepted: 26 December 2022
Published: 1 January 2023

Copyright: © 2023 by the authors. Licensee MDPI, Basel, Switzerland. This article is an open access article distributed under the terms and conditions of the Creative Commons Attribution (CC BY) license (https://creativecommons.org/licenses/by/4.0/).

1. Introduction

Imidacloprid (IMI), a neonicotinoid insecticide, is widely used for pest control in agriculture, ensuring the yield and quality of vegetables [1]. Although IMI is highly efficient for pest control, its misuse and abuse are expected to seriously threaten the ecosystem and public health [2,3]. IMI fails to degrade completely, resulting in long persistence [4]. Long-term exposure to IMI causes neurological damage, which poses a great risk to human health [5,6]. Hence, a sensitive analysis of IMI residues in foods is required to reduce health and environmental risks from hazardous materials.

Many instrumental methods, such as liquid chromatography (LC) [7], liquid chromatography–tandem mass spectrometry (LC—MS/MS) [8], and gas chromatography–tandem mass spectrometry (GC—MS/MS) [9], have been used to determine IMI residues. However, these methods are costly, complex, and time-consuming [10]. Rapid detection methods (such as immunoassays [11,12], aptamer methods [13], and electrochemical sensor methods [14,15]) with high sensitivity, simple operation, and low cost have been developed to overcome the shortcomings of these methods. In immunoassays, antibodies have the property of highly sensitive molecular recognition [16]. Enzyme-linked immunoassay (ELISA), as one of the immunoassays for the rapid evaluation of neonicotinoid insecticides, consists of two main formats: direct competitive ELISA (DC-ELISA) and indirect competitive ELISA

(IC-ELISA) [17–19]. IC-ELISA is widely used to detect pesticide residues because of its advantages, such as high sensitivity, simplicity, simple pretreatment, and high throughput detection. For instance, Zhang et al. [20] used the colorimetric (Co) IC-ELISA method to determine IMI residues with an LOD of 0.025 mg/L, showing high sensitivity and stability. Yue et al. [21] also used Co-IC-ELISA to determine organophosphate pesticides in agricultural products, obtaining an LOD for methyl parathion of 1.94 ng/mL. The method yielded favorable recoveries of 84.16–106.96% at a low spiking level. Compared to colorimetric assays, chemiluminescence, as a highly sensitive detection method, is also used to rapidly detect IMI. For instance, Girotti et al. [22] used chemiluminescence to detect IMI in honey using IC-ELISA. They found that the LOD was 0.11 ng/mL. Similarly, Hu et al. [23] detected IMI based on IC-ELISA using chemiluminescence and obtained a low LOD of 0.637 ng/mL. Colorimetric and chemiluminescent assays based on IC-ELISA have been popularly used to detect IMI with low LOD. However, the LOD of the related IC-ELISA method can be further improved.

To improve the LOD of IC-ELISA, this study reports colorimetric and chemiluminescent IC-ELISAs for detecting IMI (Figure 1). Comparing the sensitivity, linear range, antigen–antibody ratio, and detection time of Cl-ELISA and Co-ELISA provided an approach for selecting a detection assay. Finally, different sample preparations were developed for Co-ELISA and Cl-ELISA, which further improved the sensitivity and recovery of Co-ELISA and Cl-ELISA in vegetable samples, providing a reference for applying ELISAs.

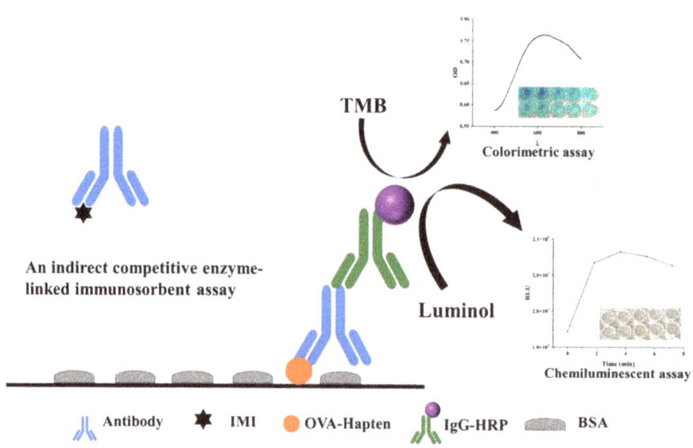

Figure 1. Illustration of the principle of Co-ELISA and Cl-ELISA.

2. Materials and Methods

2.1. Chemicals and Reagents

Imidacloprid monoclonal antibody and ovalbumin-coated hapten (OVA-Hapten) were provided by Zhejiang University (Hangzhou, China). Peroxidase-conjugated AffiniPure goat antimouse IgG was purchased from Jackson ImmunoResearch Laboratories Inc. (West Grove, PA, USA). Imidacloprid standard (100 mg/kg) was procured from Beijing Manhage Bio-Technology Co., Ltd. (Beijing, China). Acetonitrile (chromatographic grade) was procured from Merck AG (Darmstadt, Germany). TMB (3,3′,5,5′-Tetramethylbenzidine) single-component substrate solution, phosphate buffer saline tablets, and tris (hydroxymethyl) aminomethane were secured from Beijing Solarbio Science & Technology Co., Ltd. (Beijing, China). HRP-chemiluminescent reagents A and B were supplied by Beijing Keyuebio Technology Co., Ltd. (Beijing, China). Albumin (bovine serum) was obtained from Shanghai Yuanye Biotechnology Co., Ltd. (Shanghai, China). Tween 20 was acquired from Shanghai Macklin Biochemical Co., Ltd. (Shanghai, China). A 96-well opaque assay plate was obtained from Shanghai Jing An Biological Technology Co., Ltd. (Shanghai, China).

An immuno clear flat-bottom 96-well plate (item #439454#) was supplied by Thermo Fisher Scientific (Pittsburgh, PA, USA).

2.2. Optimization Parameters of the Assays

The experimental parameters (concentrations of antibody and antigen, concentration of BSA, organic solvent, and reaction time of chemiluminescence) were evaluated to improve the sensitivities of the Cl-ELISA. Seven different combinations of antigen–antibody concentrations were screened (Table 1), and the corresponding IMI calibration curves were established to obtain the best combination of antigen–antibody concentrations. When the RLU_{max}/IC_{50} was greater than 5×10^5, the minimum amount of antigen–antibody was desirable. The Cl-ELISA detected IMI under various concentrations of BSA (0.5, 1, 2, 3, and 5%), implementing methanol to the final volume (1, 5, 10, 20, and 30%) in the IMI dilution, and chemiluminescence reaction times (from 0 to 20 min). The negative well RLU value, positive well RLU value, and N/P value (negative well RLU value/positive well RLU value) were used as evaluation criteria in the optimization scheme to select the best physicochemical parameters. For Co-ELISA, the optimal parameters of the Co-ELISA protocol were screened according to the chessboard titrations, including OD = 1 and cost savings [24]. The optimal combination was selected by comparing 32 combinations of antigen–antibody concentrations (Table 2).

Table 1. The optimal concentrations of antigen and antibody for Cl-ELISA.

Antibody Concentration (mg/L)	OVA-Hapten Concentration (mg/L)			
	10		5	
	RLU_{max}/IC_{50}	R^2	RLU_{max}/IC_{50}	R^2
4	2.15×10^6	0.9676	-	-
2	7.56×10^5	0.9491	-	-
1	8.47×10^6	0.9266	1.49×10^6	0.9307
0.5	4.19×10^6	0.9582	6.24×10^6	0.9960
0.1	-	-	2.31×10^6	0.8743

Table 2. The optimal concentrations of antigen and antibody for Co-ELISA.

Antibody Dilution Times	OVA-Hapten Concentration (mg/L)			
	20	10	5	2.5
2	1.8459	1.1196	0.7011	0.5497
4	1.1879	0.8834	0.6110	0.5007
8	0.7198	0.6742	0.6787	0.5687
16	0.6613	0.6390	0.5700	0.4688
32	0.4950	0.6464	0.6006	0.5197
64	0.5268	0.4675	0.5115	0.5437
128	0.3867	0.4048	0.4691	0.4861
256	0.4395	0.4617	0.5157	0.3846

2.3. Sample Preparation

Method 1: Chinese cabbage, cucumber, and zucchini (purchased from local supermarkets) were selected as spiked samples. IMI standard concentrations (10, 50, and 100 μg/L) were used as fortification levels in homogenized samples (5 g) in 10 mL centrifuge tubes. The mixtures were stirred with 5 mL of methanol with manual shaking for 10 s and then left to extract for 30 min, followed by centrifugation at 8824 × *g* (4 °C) for 15 min. Afterwards, 3 mL of supernatant was transferred to a 10 mL plastic centrifuge tube. Finally, 100 μL of supernatant was transferred to a 2 mL plastic tube, and 900 μL of PBS buffer solution (10 mM, pH = 4) was added to ensure a 10% ratio of methanol.

Method 2: Chinese cabbage, cucumber, and zucchini squash (purchased from local supermarkets) were selected as spiked samples. IMI standard concentrations (5, 10, and 20 μg/L) were used as spiking levels for homogenized samples (5 g) in 10 mL centrifuge tubes. The mixtures were stirred with 5 mL of methanol with manual shaking for 10 s and then left to extract for 30 min, followed by centrifugation at 8824 × *g* (4 °C) for 15 min. The entire supernatant was transferred to a 15 mL centrifuge tube, and water was added to adjust the volume to 10 mL. Finally, 200 μL of the supernatant was transferred to a 2 mL plastic tube, and 800 μL of PBS buffer solution (10 mM, pH = 4) was added to ensure a 10% ratio of methanol.

2.4. Co-ELISA Establishment

A 100 μL/well OVA-Hapten (10 mg/L) in PBS (10 mM, pH 7.4) was coated on an immuno clear flat-bottom 96-well plate for 2 h. The plates were then washed three times using PBST, followed by adding a BSA (300 μL/well) blocker for 1 h. After washing three times with PBST, IMI (50 μL/well) dissolved in 10% MeOH−PBS (10 mM, pH 7.4) and antibody (2 mg/L, 50 μL/well) diluted with tris (50 mM, pH 7.4) were added and incubated for 2 h on a blocked plate. After washing, 100 μL/well of diluted (1/1000) goat antimouse IgG-HRP was added. The mixtures were incubated for 1 h, followed by the addition of TMB (100 μL/well). After incubation for 10 min, the absorbance (A_{650nm}) was read with a multifunctional microplate reader (Tecan, Salzburg, Austria). All incubations were carried out on a thermomixer (Eppendorf, Hamburg, Germany) at 37 °C and 400 rpm, followed by washing three times with PBST (10 mM PBS containing 0.1% Tween 20, pH 7.4) using a DEM-3 washing machine (Tuopu, Beijing, China).

2.5. Cl-ELISA Establishment

A 100 μL/well OVA-Hapten (5 mg/L) in PBS (10 mM, pH 7.4) was coated on 96-well opaque assay plates for 2 h. The plates were then washed, followed by the addition and incubation of BSA (300 μL/well) blocker for 1 h. After washing, analyte (50 μL/well) dissolved in 10% MeOH−PBS (10 mM, pH 7.4) and antibody (0.5 mg/L, 50 μL/well) diluted with tris (50 mM, pH 7.4) were added and incubated for 2 h on a blocked plate. After washing 3 times, 100 μL/well of diluted (1/1000) goat antimouse IgG-HRP was added. The mixtures were incubated for 1 h, followed by the addition of HRP-chemiluminescent reagents A and B (50 μL/well). After incubation for 3.6 min, the luminescence (total luminescence) was read with a multifunctional microplate reader (Tecan, Salzburg, Austria). All incubations were carried out on a thermomixer (Eppendorf, Hamburg, Germany) at 37 °C and 400 rpm, followed by washing three times with PBST (10 mM PBS containing 0.1% Tween 20, pH 7.4) using a DEM-3 washing machine (Tuopu, Beijing, China).

2.6. Calibration Curves

The calibration curve for Cl-ELISA was plotted by considering IMI concentrations on the x-axis against percent inhibition on the y-axis. The percent inhibition was calculated using the following Equation (1) [25,26]. Similarly, the Co-ELISA was calculated using Equation (2).

$$I\% = \left(1 - \frac{RLU_x - RLU_{min}}{RLU_{max} - RLU_{min}}\right) \times 100\% \quad (1)$$

$$I\% = \left(1 - \frac{OD_x - OD_{min}}{OD_{max} - OD_{min}}\right) \times 100\% \quad (2)$$

In Equation (1), RLU_{max} and RLU_x are the response without IMI and the response value when the concentration of the standard solution is x, respectively. RLU_{min} is the response value of the blank control.

In Equation (2), OD_{max} and OD_x are the response without IMI and the response value when the concentration of the standard solution is x, respectively. OD_{min} is the response value of the blank control.

3. Results

3.1. Optimization Parameters of the Assays

3.1.1. Antigen and Antibody Concentrations

The parameters of the Cl-ELISA protocol were optimized according to the chessboard assay, including signal values greater than 5.0×10^5, high RLU_{max}/IC_{50}, significant pesticide inhibition effects, and cost savings [24,26]. For Cl-ELISA, the maximum value of RLU_{max}/IC_{50} means a larger detection range and the highest sensitivity for chemiluminescence [27]. As shown in Table 1, a combination of 5 mg/L coated antigen and 0.5 mg/L antibodies were selected in the Cl-ELISA system because R^2 = 0.996, RLU_{max} 6.24×10^6 was greater than 5.0×10^5, and cost savings. As shown in Table 1, antibody concentrations that are too high or too low at constant OVA-Hapten concentrations can decrease RLU and lower sensitivity, possibly due to the hook effect [28].

For Co-ELISA, we used a chessboard assay to screen for a combination of antigen–antibody concentrations. OD = 1 is used as a target because it can save the amount of antigen and antibody used and meet the method's sensitivity [29,30]. A combination with OD = 1 and the smallest antigen–antibody concentration was used as the optimized concentration. As shown in Table 2, the higher the antigen and antibody concentrations were, the greater the OD value. When the antigen concentration was constant, the OD value decreased with increasing antibody dilution (antibody dilutions from 2 to 16 times); the OD value changed without an obvious pattern with increasing antibody dilution (antibody dilutions from 32 to 128 times). Because the concentration of antibodies is too low, the binding rate of antibodies and antigens may be reduced. The combinations of antigens with an antibody with OD values of approximately 1.0 were 20 mg/L with 4 mg/L and 10 mg/L with 2 mg/L, and the combination of 10 mg/L with 2 mg/L was chosen according to cost savings.

3.1.2. Blocking Agent

As a critical step, protein antibodies or antigens are immobilized on plastic surfaces through nonspecific binding in ELISA. Binding nonspecific proteins may lead to a decline in sensitivity and specificity and can also produce false negative results [25]. To avoid nonspecific binding, proteins were used to block the vacant sites in the plastic well. As one of the most commonly used blocking agents, BSA effectively showed optimum results in ELISA [31,32]. Therefore, BSA was chosen as the blocking solution and tested in the 0.5–5% range. According to the results (Figure 2), with increasing BSA concentration, the RLU values of the negative wells decreased continuously. The results showed that as the concentration of BSA increased, the excessive BSA adsorption on the plastic plates hindered the binding of antigens and antibodies. With increasing BSA concentration, as shown in Figure 2, the RLU values of the positive wells decreased and remained unchanged. The results suggest that the low concentration of BSA could not wholly block the adsorption sites of the plastic plates, which quickly produced false negatives. When the N/P value is the maximum, its chemiluminescence value is larger, and the sensitivity is higher. A concentration of 2% BSA was chosen as the blocking concentration.

3.1.3. Methanol

IMI solubility in water is less than that in organic solvents [33], and adding an appropriate amount of organic solvent to the buffer solution assists in the dissolution of IMI. However, an excessively high solvent content might affect the antigen–antibody reaction [34]. To compare the effect of different methanol contents on Cl-ELISA, PBS buffers containing 1%, 5%, 10%, 20%, and 30% volumes of methanol were prepared to dilute the standard solutions. The results (Figure 3) show that the RLU of the negative wells decreased with increasing methanol volume from 5% to 30%. This indicates that a high methanol volume affects antigen–antibody reactions. When the volume of methanol was 10%, the chemical RLU of the positive wells was the smallest, and the value of N/P was the largest at 4.6. The PBS buffer containing 10% methanol was selected for the Cl-ELISA method to prepare the standard pesticide solution under the conditions of pesticide solubility, maximum signal value, and detection sensitivity.

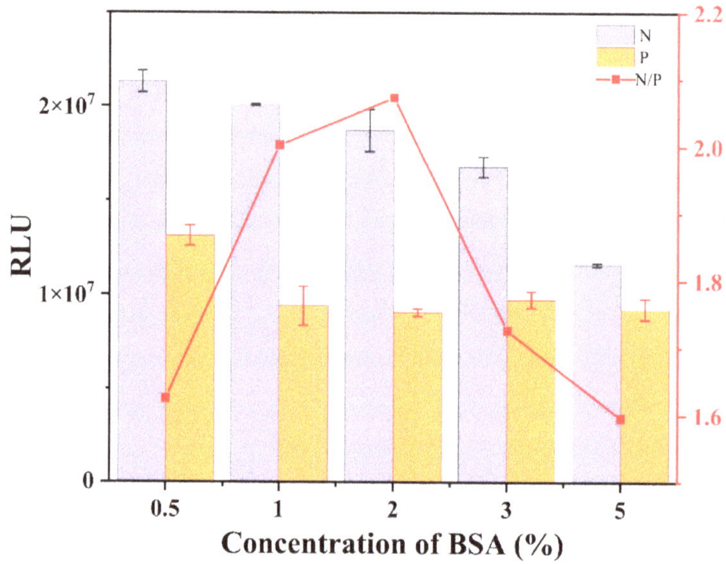

Figure 2. Effects of BSA on CI-ELISA. The data are averages of three replicates. P: the RLU values of the positive wells (the concentration of IMI = 3.125 μg/L), N: the RLU values of the negative wells (the concentration of IMI = 0).

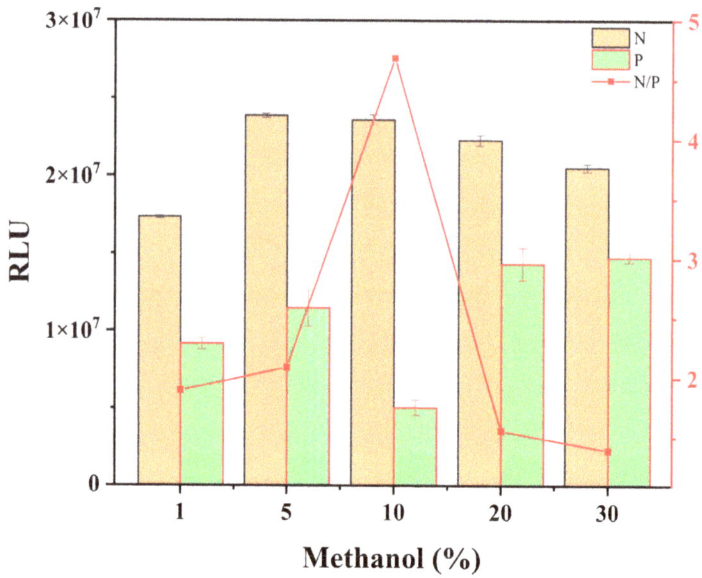

Figure 3. Effects of methanol on CI-ELISA. The data are averages of three replicates. P: the RLU values of the positive wells (the concentration of IMI = 3.125 μg/L), N: the RLU values of the negative wells (the concentration of IMI = 0).

3.1.4. Reaction Time

The substrate action time is an important factor affecting the sensitivity of Cl-ELISA. A reaction time that is too short or too long might result in a false negative. The chemiluminescence dynamic curves of the 0-well plates were analyzed under three different matrices to control the reaction time. The results (Figure 4a) showed that the RLU tended to increase and decrease with time. The plateau stabilization period of chemiluminescence (1.8–5.4 min) was approximately 3.6 min. At 3.6 min, the chemiluminescence values reached a maximum in the three matrices: cabbage 2.5×10^7, cucumber 2.0×10^7, and zucchini 1.7×10^7.

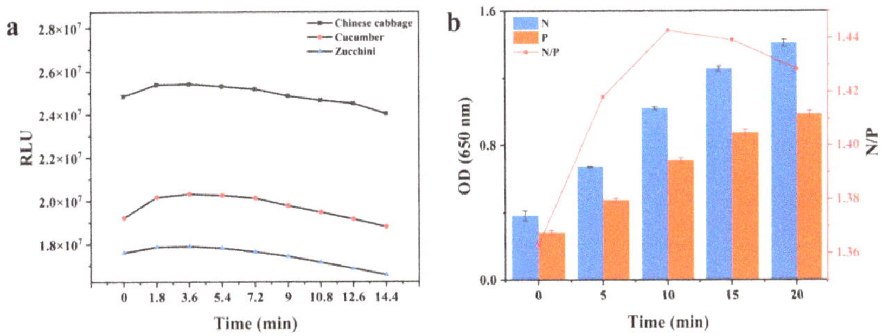

Figure 4. (**a**) The dynamic curve of Cl-ELISA in the three tested vegetable samples from 0 to 14.4 min with no analyte. (**b**) Absorbance values for optimization of reaction time (0, 5, 10, 15, and 20 min) for Co-ELISA.

Under optimal conditions, Co-ELISA was set at 37 °C with substrate action times of 0, 5, 10, 15, and 20 min (Figure 4b). With increasing time, the OD values of the negative and positive wells increased. At 10 min, the OD value was approximately 1.0, and the N/P value was a maximum. Thus, 10 min was chosen as the best action time for the TMB solution. The time of the maximum N/P value was chosen as the optimal action time (10 min) of the TMB solution. The RLU of Cl-ELISA could be measured immediately after the addition of the HRP-chemiluminescent reagents (3.6 min), while the colorimetric method required 10 min of reaction.

3.2. Calibration Curves

All these factors were accounted for, and the optimal conditions for Cl-ELISA were as follows: coated antigen (5 mg/L) and antibody (0.5 mg/L) produced the highest RLU_{max}/IC_{50} ratio; 10% methanol, 2% BSA, and 3 min substrate action time were used for Cl-ELISA. Under optimal conditions, Figure 5a shows the Cl-ELISA standard curve for IMI. A calibration curve was obtained based on a similar linear section of the standard curve (inset of Figure 5a). The limit of detection (LOD) and the sensitivity (IC_{50}) of Cl-ELISA were 0.19 μg/L and 2.66 μg/L, respectively. Similarly, the optimal conditions for Co-ELISA were as follows: coated antigen (10 mg/L) and antibody (2 mg/L) produced the OD = 1; 10% methanol, 2% BSA, and 10 min substrate action time were used for Co-ELISA. Under optimal conditions, Figure 5b shows the Co-ELISA standard curve for IMI. A calibration curve was obtained based on a similar linear section of the standard curve (inset of Figure 5b). The limit of detection (LOD) and the sensitivity (IC_{50}) of Cl-ELISA were 1.56 μg/L and 8.15 μg/L, respectively.

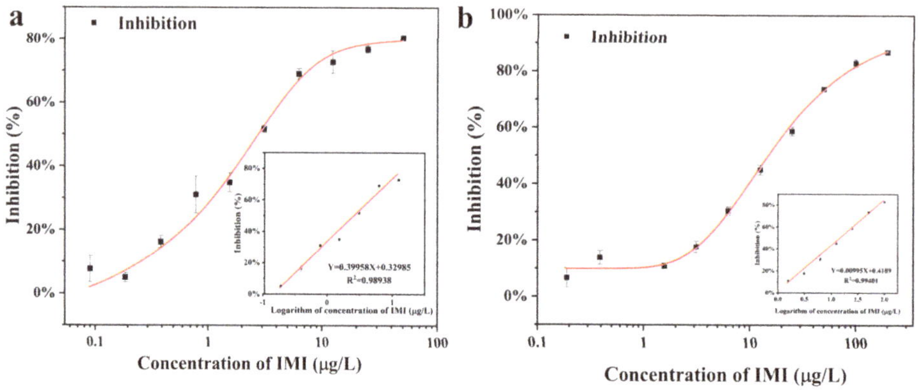

Figure 5. (**a**) Standard curve for the detection of IMI using Cl-ELISA. (**b**) Standard curve for the detection of IMI using Co-ELISA.

Figure 5 shows the Co-ELISA linearity over a wide range (1.56–200 μg/L) with R^2 = 0.9893. In comparison, the Cl-ELISA linearity range (0.19–25 μg/L) with R^2 = 0.9940 may be caused by the concentration of antigen and antibody of Co-ELISA being greater than that of Cl-ELISA. The sensitivity of Cl-ELISA (IC_{50} = 1.56 μg/L) was greater than that of Co-ELISA (IC_{50} = 8.15 μg/L), which may be due to the sensitivity of the chemiluminescence assay compared to that of the colorimetric assay (Table 3). From Table 3, for Co-ELISA, the linearity (1.56–200 μg/L) was broader than that of Cl-ELISA (0.19–25 μg/L), which achieved quantitative analysis in two orders of magnitude ranges. For Cl-ELISA, the concentration of antigen and antibody was economical compared to Co-ELISA. The IC_{50} (1.56 μg/L) of Cl-ELISA was more sensitive than the IC_{50} (8.15 μg/L) of Co-ELISA, which was caused by the chemiluminescence assay being more sensitive than the colorimetric assay. In addition, chemiluminescent assays were time-saving and efficient compared to colorimetric assays.

Table 3. Comparison between Cl-ELISA and Co-ELISA.

Parameters	Cl-ELISA	Co-ELISA
Antigen (mg/L)	5	10
Antibody (mg/L)	0.5	2
Linearity (μg/L)	0.19–25	1.56–200
R^2	0.9940	0.9893
Reaction time (min)	3.6	10
IC_{50} (μg/L)	1.56	8.15

3.3. *Optimization of Sample Preparation*

Method 1 was the sample preparation for Cl-ELISA and Co-ELISA. Co-ELISA showed good recovery, while Cl-ELISA had poor recovery (26.8–237.5%) and RSD (1.4–52.7%) (Table 4). The recovery and RSD of Cl-ELISA could not meet the recovery (between 60 and 120%) and RSD (≤30%), according to NY/T 788-2018 [35]. Due to the mutual solubility of methanol and water, the proportion of methanol in the extracted supernatant could not be determined, thereby, reducing the sensitive detection of trace IMI in Cl-ELISA. Thus, method 1 was optimized by transferring all the supernatant to a centrifuge tube, after which water was added to fix the volume of methanol to 10 mL to ensure a constant volume of methanol. A trace amount of IMI could be extracted completely, making Cl-ELISA obtain good recovery.

Table 4. Recovery of spiked IMI in vegetable samples (Chinese cabbage, cucumber, and zucchini) using Co-ELISA and Cl-ELISA (n = 3).

	Method 1				Method 1				Method 2			
Sample	**Spiked (μg/L)**	**Co-ELISA (μg/L)**	**Recovery (%)**	**RSD (%)**	**Spiked (μg/L)**	**Cl-ELISA (μg/L)**	**Recovery (%)**	**RSD (%)**	**Spiked (μg/L)**	**Cl-ELISA (μg/L)**	**Recovery (%)**	**RSD (%)**
Chinese cabbage	10	11.35 ± 1.87	113.5	5.4	5	1.34 ± 0.46	26.8	7.5	5	5.47 ± 0.14	109.3	4.7
	50	58.74 ± 1.94	117.5	6.2	10	5.85 ± 1.13	58.5	5.6	10	7.03 ± 0.11	70.3	1.5
	100	81.71 ± 2.02	81.7	6.9	20	18.19 ± 0.56	91.0	1.4	20	13.94 ± 0.17	69.7	8.1
cucumber	10	8.21 ± 1.14	82.1	0.7	5	5.06 ± 0.04	101.2	8.1	5	4.49 ± 0.69	89.7	7.9
	50	49.11 ± 1.20	98.2	1.7	10	7.23 ± 0.09	72.3	24.0	10	11.41 ± 0.47	114.1	1.0
	100	79.38 ± 1.31	79.4	3.2	20	9.30 ± 0.60	46.5	19.5	20	14.82 ± 0.60	74.1	5.5
zucchini	10	11.11 ± 2.05	111.1	7.9	5	9.35 ± 0.30	187.0	43.8	5	6.07 ± 0.17	120.6	5.0
	50	58.81 ± 2.42	117.6	11.1	10	23.74 ± 0.18	237.5	45.7	10	6.99 ± 0.43	69.9	15.0
	100	82.55 ± 2.50	82.5	11.6	20	27.10 ± 0.19	135.5	52.7	20	19.94 ± 0.24	104.9	8.8

3.4. Accuracy and Precision

The inhibition curves for IMI in Chinese cabbage, cucumber, and zucchini matrices using Co-ELISA and Cl-ELSIA are shown in Figures 6 and 7. Both Co-ELISA and Cl-ELISA achieved quantitative analysis of IMI with high sensitivity. The linearity of Co-ELISA was observed in the range of 3.125 μg/L–100 μg/L in Chinese cabbage (R^2 = 0.9746), 1.56 μg/L–200 μg/L in cucumber (R^2 = 0.9743), and 1.56 μg/L–200 μg/L in zucchini (R^2 = 0.9756). The linearity of Cl-ELISA was observed in the range of 0.39 μg/L–25 μg/L in Chinese cabbage with a good regression coefficient (R^2) of 0.9656, 1.56 μg/L–50 μg/L in cucumber (R^2 = 0.9577), and 0.39 μg/L–50 μg/L in zucchini (R^2 = 0.9873). Chinese cabbage, cucumber, and zucchini were spiked at concentrations of 10, 50, and 100 μg/L for Co-ELISA and concentrations of 5, 10, and 20 μg/L for Cl-ELISA. The recovery was performed under optimized conditions and compared with Cl-ELISA for its performance. According to Table 4, for Co-ELISA, the recovery of IMI in spiked samples varied from 81.7 to 117.5% in Chinese cabbage (RSD between 5.4 and 6.9%), 79.4 to 98.2% in cucumber (RSD between 0.7 and 3.2%), and 82.5 to 117.6% in zucchini (RSD between 7.9 and 11.6%). For Cl-ELISA, recovery varied between 69.7 and 109.3% in Chinese cabbage (RSD between 1.5 and 8.1%), 74.1 and 114.1% in cucumber (RSD between 1.0 and 7.9%), and 69.9 and 120.6% in zucchini (RSD between 5.0 and 15.0%). Comparing the two assays, Co-ELISA showed the best recovery results, where 79.4 to 117.6% was achieved for all samples (Table 4). At a lower concentration of 5 μg/L, Cl-ELISA showed better recovery in cucumber than in Chinese cabbage and zucchini, which could be due to matrix effects. According to China's and the EU's pesticide residue limit standards, (the MRL of IMI in Chinese cabbage is 0.2 mg/kg and that of cucumber and zucchini is 1 mg/kg, China) (the MRL of IMI in Chinese cabbage is 0.01 mg/kg and that of cucumber is 0.5 mg/kg, EU) Co-ELISA and Cl-ELISA can meet the requirements of rapidly detecting IMI. Co-ELISA and Cl-ELISA are suitable to ensure rapid, reliable, and sensitive detection of IMI in Chinese cabbage, cucumber, and zucchini samples.

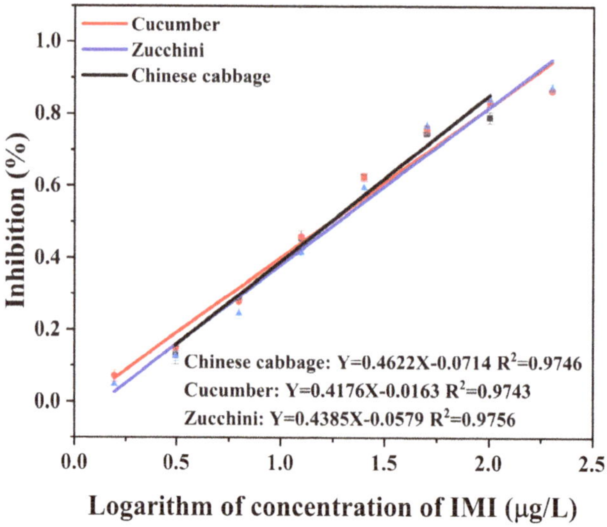

Figure 6. Standard curve for the detection of IMI in vegetable samples (Chinese cabbage, cucumber, and zucchini) using Co-ELISA.

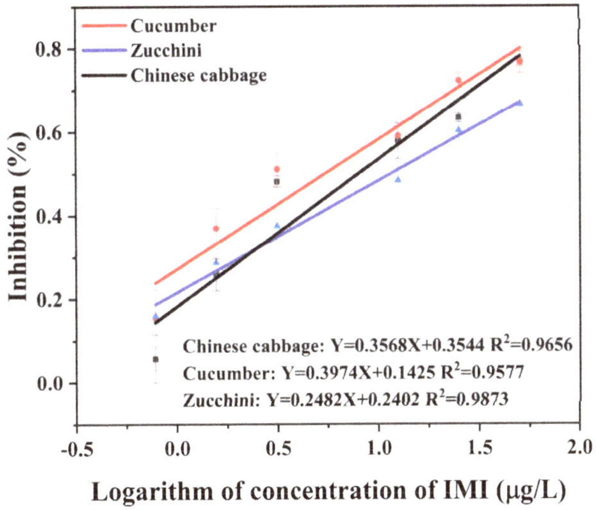

Figure 7. Standard curve for the detection of IMI in vegetable samples (Chinese cabbage, cucumber, and zucchini) using Cl-ELISA.

3.5. Comparison with Other Methods for Detecting IMI

Table 5 shows the LOD and linear range of different instruments for detecting IMI. The LOD = 1.56 μg/L of Co-ELISA and the LOD = 0.19 μg/L of Cl-ELISA were lower than those of HPLC and HPLC—MS/MS in Table 5, which proved that Co-ELISA and Cl-ELISA were feasible. Some immunoassay methods detect IMI with significantly lower LODs than this work and show better results. For example, Wang et al. [36] used an immunochromatographic method based on scandium-tetrakis (4-carboxyphenyl) porphyrin (TCPP) metal–organic framework nanocubes to detect IMI in vegetables, and the Sc-TCPP 3D MOF had good biocompatibility and optical properties, thus, enhancing the sensitivity of the detection. In addition, Fernández et al. [37] constructed an immunosensor for the detection of IMI based on nanogold electrodes and obtained a low LOD of 22 pmol/L. Guo et al. [38] established immunoassay based on graphene oxide (GO) and up-converting nanoparticles (UCNPs) showed a wide detection range of 0.08–50 ng/mL to IMI. The combination of immunoassays and nanomaterials allows for more sensitive detection.

Table 5. Comparison of instruments detecting IMI (n = 3).

Methods	LOD (μg/L)	Recovery (%)	Sample	RSD (%)	Reference
HPLC—MS/MS	5.9	70.1–119.3	fruits	2.7–12.4	[39]
HPLC	2.7	88.9–93.9	juice	3.2–4.1	[40]
HPLC	10	83.3–90.8	soil	2.4–4.4	[41]
HPLC—MS/MS	2.1	85.9–103.8	vegetable	2.6–12.1	[42]
HPLC	8.53	88.9–90.2	vegetable	5.5–6.8	[43]
ITS	0.04	80.0–124.0	vegetable	0.1–2.7	[36]
Immunoassay (UCNPs/GO)	0.08	76.8–101.8	Water, vegetable, tea, honey	-	[38]
Co-ELISA Cl-ELISA	1.56 0.19	79.4–117.6 69.7–120.6	vegetable vegetable	0.7–11.6 1.0–15.0	This work

ITS: immunochromatographic test strip; reprinted/adapted with permission from Refs. [36,38]. Copyright 2022, Elsevier; reprinted/adapted with permission from Ref. [41]. Copyright 2022, IOP science; reprinted/adapted with permission from Ref. [43]. Copyright 2022, Taylor & Francis; reprinted/adapted with permission from Chinese Refs. [39,40,42].

4. Conclusions

Herein, Co-ELISA and Cl-ELISA were developed for detecting IMI in Chinese cabbage, cucumber, and zucchini. The recoveries ranged from 81.7 to 117.6% for Co-ELISA and 69.7 to 120.6% for Cl-ELISA. The principle of detection used for both assays can also be applied to detect IMI at a sensitive level in other vegetable samples. Co-ELISA exhibited a wide linear range (1.56–200 μg/L) suitable for IMI testing at the microgram level. Compared with Co-ELISA, Cl-ELISA showed higher sensitivity (IC_{50} = 2.66 μg/L), benefiting the trace detection level, which was one order of magnitude higher than that of Co-ELISA. Cl-ELISA is a cost-saving alternative to Co-ELSIA in terms of the amount of antigen and antibody used and detection time. This study builds colorimetric and chemiluminescent assays based on ELISA detection of IMI and provides the theoretical basis for selecting an optical detection assay.

Author Contributions: Conceptualization, R.Z. and G.C.; methodology, G.L.; software, X.H.; validation, X.X., X.H. and G.C.; formal analysis, G.C.; investigation, L.L.; resources, Y.Z.; data curation, G.C.; writing—original draft preparation, R.Z. and G.C.; writing—review and editing, G.C. and A.M.A.; visualization, X.X.; supervision, D.X.; project administration, X.H.; funding acquisition, D.X. All authors have read and agreed to the published version of the manuscript.

Funding: This study was financially supported by the Agricultural Science and Technology Innovation Program of CAAS (CAAS-ZDRW202011, CAAS-TCX2019025-5), the China Agriculture Research System of MOF and MARA (CARS-23-E03), and the National Key Research Development Program of China (2020YFD1000300).

Data Availability Statement: The original contributions presented in this study are included in the article, and further inquiries can be directed to the corresponding authors.

Conflicts of Interest: The authors declare no conflict of interest.

References

1. Zhai, R.; Zhang, K.; Chen, G.; Liu, G.; Huang, X.; Gao, M.; Zhou, J.; Xu, X.; Li, L.; Zhang, Y.; et al. Residue, Dissipation Pattern, and Dietary Risk Assessment of Imidacloprid in Chinese Chives. *Front. Nutr.* **2022**, *9*, 846333. [CrossRef] [PubMed]
2. Yang, L.; Shen, Q.; Zeng, T.; Li, J.; Li, W.; Wang, Y. Enrichment of imidacloprid and its metabolites in lizards and its toxic effects on gonads. *Environ. Pollut.* **2020**, *258*, 113748. [CrossRef] [PubMed]
3. Seifrtova, M.; Halesova, T.; Sulcova, K.; Riddellova, K.; Erban, T. Distributions of imidacloprid, imidacloprid-olefin and imidacloprid-urea in green plant tissues and roots of rapeseed (*Brassica napus*) from artificially contaminated potting soil. *Pest Manag. Sci.* **2017**, *73*, 1010–1016. [CrossRef] [PubMed]
4. Van Loon, S.; Vicente, V.B.; van Gestel, C.A.M. Long-Term Effects of Imidacloprid, Thiacloprid, and Clothianidin on the Growth and Development of Eisenia andrei. *Environ. Toxicol. Chem.* **2022**, *41*, 1686–1695. [CrossRef]
5. Sriapha, C.; Trakulsrichai, S.; Tongpoo, A.; Pradoo, A.; Rittilert, P.; Wananukul, W. Acute Imidacloprid Poisoning in Thailand. *Ther. Clin. Risk Manag.* **2020**, *16*, 1081–1088. [CrossRef]
6. Du, M.; Yang, Q.; Liu, W.; Ding, Y.; Chen, H.; Hua, X.; Wang, M. Development of immunoassays with high sensitivity for detecting imidacloprid in environment and agro-products using phage-borne peptides. *Sci. Total Environ.* **2020**, *723*, 137909. [CrossRef] [PubMed]
7. Rancan, M.; Sabatini, A.G.; Achilli, G.; Galletti, G.C. Determination of Imidacloprid and metabolites by liquid chromatography with an electrochemical detector and post column photochemical reactor. *Anal. Chim. Acta* **2006**, *555*, 20–24. [CrossRef]
8. Iwafune, T.; Ogino, T.; Watanabe, E. Water-based extraction and liquid chromatography-tandem mass spectrometry analysis of neonicotinoid insecticides and their metabolites in green pepper/tomato samples. *J. Agric. Food Chem.* **2014**, *62*, 2790–2796. [CrossRef]
9. Lodevico, R.G.; Li, Q.X. Determination of Total Imidacloprid Residues in Coffee by Gas Chromatography–Mass Spectrometry. *Analytical Letters.* **2002**, *35*, 315–326. [CrossRef]
10. Zhai, R.; Chen, G.; Liu, G.; Huang, X.; Xu, X.; Li, L.; Zhang, Y.; Wang, J.; Jin, M.; Xu, D.; et al. Enzyme inhibition methods based on Au nanomaterials for rapid detection of organophosphorus pesticides in agricultural and environmental samples: A review. *J. Adv. Res.* **2022**, *37*, 61–74. [CrossRef]
11. Li, H.; Jin, R.; Kong, D.; Zhao, X.; Liu, F.; Yan, X.; Lin, Y.; Lu, G. Switchable fluorescence immunoassay using gold nanoclusters anchored cobalt oxyhydroxide composite for sensitive detection of imidacloprid. *Sens. Actuators B Chem.* **2019**, *283*, 207–214. [CrossRef]
12. Perez-Fernandez, B.; Mercader, J.V.; Checa-Orrego, B.I.; de la Escosura-Muniz, A.; Costa-Garcia, A. A monoclonal antibody-based immunosensor for the electrochemical detection of imidacloprid pesticide. *Analyst* **2019**, *144*, 2936–2941. [CrossRef] [PubMed]

13. Bor, G.; Man, E.; Ugurlu, O.; Ceylan, A.E.; Balaban, S.; Durmus, C.; Pinar Gumus, Z.; Evran, S.; Timur, S. in vitro Selection of Aptamer for Imidacloprid Recognition as Model Analyte and Construction of a Water Analysis Platform. *Electroanalysis* **2020**, *32*, 1922–1929. [CrossRef]
14. Wang, S.H.; Lo, S.C.; Tung, Y.J.; Kuo, C.W.; Tai, Y.H.; Hsieh, S.Y.; Lee, K.L.; Hsiao, S.R.; Sheen, J.F.; Hsu, J.C.; et al. Multichannel nanoplasmonic platform for imidacloprid and fipronil residues rapid screen detection. *Biosens. Bioelectron.* **2020**, *170*, 112677. [CrossRef]
15. Zhang, W.; Liu, C.; Han, K.; Wei, X.; Xu, Y.; Zou, X.; Zhang, H.; Chen, Z. A signal on-off ratiometric electrochemical sensor coupled with a molecular imprinted polymer for selective and stable determination of imidacloprid. *Biosens. Bioelectron.* **2020**, *154*, 112091. [CrossRef]
16. Li, H.; He, S.; Liu, G.; Li, C.; Ma, Z.; Zhang, X. Residue and dissipation kinetics of toosendanin in cabbage, tobacco and soil using IC-ELISA detection. *Food Chem.* **2021**, *335*, 127600. [CrossRef]
17. Watanabe, E.; Miyake, S.; Yogo, Y. Review of enzyme-linked immunosorbent assays (ELISAs) for analyses of neonicotinoid insecticides in agro-environments. *J. Agric. Food Chem.* **2013**, *61*, 12459–12472. [CrossRef]
18. Zhang, Z.; Chen, Q.; Huang, H.; Zhang, K.; Bai, L.; Tan, G. Ultrasensitive Immunoassay for the Determination of Imidacloprid in Medicinal Herbs. *Anal. Lett.* **2022**, 1–13. [CrossRef]
19. Watanabe, E.; Baba, K.; Eun, H.; Miyake, S. Application of a commercial immunoassay to the direct determination of insecticide imidacloprid in fruit juices. *Food Chem.* **2007**, *102*, 745–750. [CrossRef]
20. Zhang, L.; Wang, Z.; Wen, Y.; Shi, J.; Wang, J. Simultaneous detection of parathion and imidacloprid using broad-specificity polyclonal antibody in enzyme-linked immunosorbent assay. *Anal. Methods* **2015**, *7*, 205–210. [CrossRef]
21. Yue, Y.; Chen, J.; Zhang, M.; Yin, Y.; Dong, Y. Determination of Organophosphorus Pesticides in Vegetables and Fruit by an Indirect Competitive Enzyme-Linked Immunosorbent Assay (ic-ELISA) and a Lateral-Flow Immunochromatographic (LFIC) Strip Assay. *Anal. Lett.* **2022**, *55*, 1701–1718. [CrossRef]
22. Girotti, S.; Maiolini, E.; Ghini, S.; Eremin, S.; Mañes, J. Quantification of Imidacloprid in Honeybees: Development of a Chemiluminescent ELISA. *Anal. Lett.* **2010**, *43*, 466–475. [CrossRef]
23. Hu, Y.M.; Feng, H.W.; Liu, S.; Liu, C.; Zhao, P.Y.; Zhang, M.; Zhang, L.; Zhao, J.; Li, J.Z.; Yu, X.M.; et al. The preparation of polyclonal antibody against chlordimeform and establishment of detection by indirect competitive ELISA. *J. Environ. Sci. Health B* **2022**, *57*, 114–124. [CrossRef] [PubMed]
24. Tan, W.M.; He, S.P.; Zhang, L.; Zhao, H.W.; Zhao, J.; Li, Z.; Li, X.F.; Wang, B.M. Systematic optimization of antibody and coating antigen concentrations in enzyme linked immunosorbent assay checkerboard assay. *Chin. J. Anal. Chem.* **2008**, *36*, 1191–1195.
25. Mukherjee, M.; Nandhini, C.; Bhatt, P. Colorimetric and chemiluminescence based enzyme linked apta-sorbent assay (ELASA) for ochratoxin A detection. *Spectrochim. Acta A Mol. Biomol. Spectrosc.* **2021**, *244*, 118875. [CrossRef]
26. Zou, R.B.; Liu, Y.; Wang, S.J.; Zhang, Y.; Guo, Y.R.; Zhu, G.N. Development and evaluation of chemiluminescence enzyme-linked immunoassay for residue detection of three organophosphorus pesticides. *Chin. J. Pestic. Sci.* **2017**, *19*, 37–45.
27. Ding, Y.; Hua, X.; Sun, N.; Yang, J.; Deng, J.; Shi, H.; Wang, M. Development of a phage chemiluminescent enzyme immunoassay with high sensitivity for the determination of imidaclothiz in agricultural and environmental samples. *Sci. Total Environ.* **2017**, *609*, 854–860. [CrossRef]
28. Shin, S.; Choi, M.; Shim, J.; Park, S. Hook effect detection and detection-range-controllable one-step immunosensor for inflammation monitoring. *Sens. Actuators B Chem.* **2020**, *304*, 127408. [CrossRef]
29. Song, G.; Huang, L.; Huang, Y.; Liu, W.; Wang, M.; Hua, X. Electrofusion preparation of anti-triazophos monoclonal antibodies for development of an indirect competitive enzyme-linked immunosorbent assay. *J. Immunol. Methods* **2022**, *500*, 113184. [CrossRef]
30. Parra, J.; Mercader, J.V.; Agullo, C.; Abad-Somovilla, A.; Abad-Fuentes, A. Generation of anti-azoxystrobin monoclonal antibodies from regioisomeric haptens functionalized at selected sites and development of indirect competitive immunoassays. *Anal. Chim. Acta* **2012**, *715*, 105–112. [CrossRef]
31. Ma, H.; Xu, Y.; Li, Q.X.; Xu, T.; Wang, X.; Li, J. Application of enzyme-linked immunosorbent assay for quantification of the insecticides imidacloprid and thiamethoxam in honey samples. *Food Addit. Contam: Part A.* **2009**, *26*, 713–718. [CrossRef]
32. Liu, Z.; Liu, J.; Wang, K.; Li, W.; Shelver, W.L.; Li, Q.X.; Li, J.; Xu, T. Selection of phage-displayed peptides for the detection of imidacloprid in water and soil. *Anal. Biochem.* **2015**, *485*, 28–33. [CrossRef] [PubMed]
33. Kong, M.Z.; Shi, X.H.; Cao, Y.C.; Zhou, C.R. Solubility of Imidacloprid in Different Solvents. *J. Chem. Eng. Data* **2008**, *53*, 615–618. [CrossRef]
34. Gui, W.J.; Liu, Y.H.; Wang, C.M.; Liang, X.; Zhu, G.N. Development of a direct competitive enzyme-linked immunosorbent assay for parathion residue in food samples. *Anal. Biochem.* **2009**, *393*, 88–94. [CrossRef] [PubMed]
35. Fu, Y.; Wang, Q.; Zhang, L.; Ling, S.; Jia, H.; Wu, Y. Dissipation, occurrence, and risk assessment of 12 pesticides in Dendrobium officinale Kimura et Migo. *Ecotoxicol. Environ. Saf.* **2021**, *222*, 112487. [CrossRef] [PubMed]
36. Wang, Y.; Zhang, M.; Bu, T.; Bai, F.; Zhao, S.; Cao, Y.; He, K.; Wu, H.; Xi, J.; Wang, L. Immunochromatographic Assay based on Sc-TCPP 3D MOF for the rapid detection of imidacloprid in food samples. *Food Chem.* **2023**, *401*, 134131. [CrossRef] [PubMed]
37. Perez-Fernandez, B.; Mercader, J.V.; Abad-Fuentes, A.; Checa-Orrego, B.I.; Costa-Garcia, A.; Escosura-Muniz, A. Direct competitive immunosensor for Imidacloprid pesticide detection on gold nanoparticle-modified electrodes. *Talanta* **2020**, *209*, 120465. [CrossRef]

38. Guo, Y.; Zou, R.; Si, F.; Liang, W.; Zhang, T.; Chang, Y.; Qiao, X.; Zhao, J. A sensitive immunoassay based on fluorescence resonance energy transfer from up-converting nanoparticles and graphene oxide for one-step detection of imidacloprid. *Food Chem.* **2021**, *335*, 127609. [CrossRef] [PubMed]
39. Liang, X.; Zhang, W.Y.; Zhang, W. Simultaneous Determination of Residues of 38 Pesticides in Fruits by QuEChERS Combined with High Performance Liquid Chromatography-Tandem Mass Spectrometry. *Food Sci.* **2020**, *41*, 288–296.
40. Tursen, ·J.; Yang, T.; Bai, L.; Tan, R. Determination of imidacloprid and acetamiprid in grenadine juice by vortex-assisted dispersive liquid-liquid microextraction coupled with HPLC. *Chin. J. Anal. Lab.* **2019**, *38*, 815–818.
41. Xue, Q.H.; Huang, Q.; Sun, X.H. Effects of Spent Mushroom Compost on Degradation of Imidacloprid in Soil and Plant. *IOP Conf. Ser. Earth Environ. Sci.* **2019**, *252*, 052042. [CrossRef]
42. Liu, J.G.; Sun, H.; Cheng, X.; Yang, J.; Hou, Z.G. Determination of Paclobutrazol and Other Four Pesticides in the Vegetable by HPLC-MS/MS. *Agrochemicals* **2012**, *51*, 358–363.
43. Badawy, M.E.I.; Ismail, A.M.E.; Ibrahim, A.I.H. Quantitative analysis of acetamiprid and imidacloprid residues in tomato fruits under greenhouse conditions. *J. Environ. Sci. Health B* **2019**, *54*, 898–905. [CrossRef] [PubMed]

Disclaimer/Publisher's Note: The statements, opinions and data contained in all publications are solely those of the individual author(s) and contributor(s) and not of MDPI and/or the editor(s). MDPI and/or the editor(s) disclaim responsibility for any injury to people or property resulting from any ideas, methods, instructions or products referred to in the content.

foods

MDPI

Article

Development of New Antibodies and an ELISA System to Detect the Potato Alkaloids α-Solanine and α-Chaconine

Kohki Okada [1,]* and Kano Matsuo [2]

1 Department of Medical Technology and Sciences, Faculty of Health Sciences, Kyoto Tachibana University, Kyoto 607-8175, Japan
2 Graduate School of Health Sciences, Kyoto Tachibana University, Kyoto 607-8175, Japan
* Correspondence: okada-ko@tachibana-u.ac.jp

Abstract: Food poisoning can be caused by the potato alkaloids α-solanine (SO) and α-chaconine (CHA). Therefore, this study aimed to establish new enzyme-linked immunosorbent assays (ELISAs) for detecting these two toxins in biological samples and potato extracts. Two antibodies that bind to solanidine, a chemical compound found in both SO and CHA, were newly developed, and two types of ELISAs (Sold1 ELISA and Sold2 ELISA) were constructed. We measured SO and CHA diluted in phosphate-buffered saline (PBS), serum, and urine. The detection performance of the two ELISAs for SO and CHA in PBS was higher than in serum and urine, and the sensitivity of Sold2 ELISA was lower than that of Sold1 ELISA. Thus, we used these ELISAs to measure SO and CHA in potato part extracts and found that potato sprouts contained approximately 80-fold more SO and CHA than tubers and 8-fold more SO and CHA than peels. Although the detection sensitivity of SO and CHA depends on the sample types, these ELISAs may be effective as future clinical and food testing methods after further improvements.

Keywords: α-chaconine; α-solanine; enzyme-linked immunosorbent assay; food poisoning; food testing; glycoalkaloid; laboratory medicine; potato alkaloids; solanidine

Citation: Okada, K.; Matsuo, K. Development of New Antibodies and an ELISA System to Detect the Potato Alkaloids α-Solanine and α-Chaconine. *Foods* **2023**, *12*, 1621. https://doi.org/10.3390/foods12081621

Academic Editor: Maojun Jin

Received: 10 March 2023
Revised: 7 April 2023
Accepted: 7 April 2023
Published: 12 April 2023

Copyright: © 2023 by the authors. Licensee MDPI, Basel, Switzerland. This article is an open access article distributed under the terms and conditions of the Creative Commons Attribution (CC BY) license (https://creativecommons.org/licenses/by/4.0/).

1. Introduction

α-solanine (SO) and α-chaconine (CHA) are natural toxins that are mainly produced in potatoes in response to pest attacks [1] and comprise approximately 95% of the total glycoalkaloid (GA) content in potatoes. The oral ingestion of these toxins often causes food poisoning in humans [2]; according to a previous report, ingestion of more than 1 mg of GAs per kg of body weight is toxic to humans [3]. Symptoms of GA food poisoning include vomiting, diarrhea, cardiac dysrhythmia, and inflammation and pain in the joints [4]. Although the concentration of GAs varies among potato species, it generally ranges from 0.4 to 1000 μg/g in commercial potatoes [5]. GAs are more abundant in the peel and sprout of potatoes than in the tuber, and light exposure and long-term storage further increase the GA content in these parts [6–8]. Furthermore, only a few percent of these toxins are eliminated during cooking, such as boiling or frying [7]. Given the global distribution of potatoes, the occurrence of GA food poisoning continues to be inevitable, however, efforts should be taken to reduce the number of fatal patients.

The development of food testing methods is important to avoid the consumption of potatoes rich in GAs. In 1994, Stanker et al. succeeded in developing the first monoclonal antibody against GAs [9]. The limit of detection (LOD) for GAs measured using an enzyme-linked immunosorbent assay (ELISA) kit constructed with this antibody was approximately 70 ng/mL [10]. Simultaneously, the measurement of GAs in potatoes using high-performance liquid chromatography (HPLC) was developed, and the LODs of SO and CHA calculated using this method were 1.2 ng/mL and 1.3 ng/mL, respectively [8,11]. Recently, the detection performance of GAs in potatoes has been significantly improved using the combination of HPLC and mass spectrometry [12]. Although several methods for detecting GAs in potato extracts have been reported, few studies have detected GAs in biological samples, such as serum and urine, which is critical for diagnosing patients with GA food poisoning. Hellenäs et al. were the first to successfully detect potato GAs in serum at a LOD of 0.3 ng/mL using HPLC [13]. Mensinga et al. also used HPLC to measure the changes in human serum GA concentrations over time, which were detectable in the range of 0.5–50 ng/mL [14]. Despite several components in serum, HPLC can efficiently detect GAs in serum. However, in recent years, few studies have developed an efficient detection method for potato-derived SO and CHA in biological samples.

The effects of eating stale potatoes should not be underestimated, as there have been some cases of deaths due to GA food poisoning [15]. However, in previous case reports, the diagnosis of potato food poisoning was based on the fact that the patients consumed old potatoes and suffered from clinical symptoms and not on the measurement of GAs in their biological samples [16]. Few medical institutions diagnose potato food poisoning by measuring GAs. As many hospitals are not equipped with HPLC or mass spectrometry instruments, it is difficult to effectively and rapidly measure the GA contents in the biological samples of patients with GA food poisoning. If there is an established laboratory method for measuring GAs in biological samples, physicians or medical practitioners can determine whether the treatment of the patients is appropriate and whether any GAs are retained in the body. In order to develop a laboratory method for detecting GAs, it is important to generate antibodies that can reliably capture GAs. By using superior antibodies as reagents, immunological assays can be constructed or incorporated into automated analyzers in medical facilities.

In the present study, we aimed to identify new antibodies that can bind to both SO and CHA and construct two types of ELISAs using them. Further, we evaluated the detection performance of these ELISAs for SO and CHA in two biological samples (serum and urine) and potato part extracts to verify their usefulness as clinical and food testing methods. We found that although the ELISAs established in our study had low detection sensitivity for the serum and urine samples, they can be used as clinical diagnosis methods after certain improvements. Further, as they can efficiently measure the SO and CHO concentrations in potato extracts, they also have the potential to be used as food testing methods.

2. Materials and Methods

2.1. Reagents

SO and CHA powders were purchased from Sigma-Aldrich Co., LLC (Tokyo, Japan). Horseradish peroxidase (HRP)-labeled goat anti-rabbit polyclonal IgG (H + L) antibodies were obtained from Funakoshi Co., Ltd. (Tokyo, Japan). Commercial serum and urine samples of healthy volunteers were obtained from Cosmo Bio Co., Ltd. (Tokyo, Japan). Bio-Safe Coomassie stain, Clarity Western ECL substrate kits, Precision Plus Protein Dual Color Standards, Trans-Blot Transfer Packs, and 0.2 µm pore-size nitrocellulose membranes were obtained from Bio-Rad Laboratories, Inc. (Hercules, CA, USA). A 96-well ELISA plate H was purchased from Sumitomo Bakelite Co., Ltd. (Tokyo, Japan). An HRP labeling kit was obtained from Dojindo Laboratories (Kumamoto, Japan). Blocking One and dimethyl sulfoxide (DMSO) were purchased from Nacalai Tesque, Inc (Kyoto, Japan). The Prominence LC-2010 HPLC system and the reversed-phase chromatography (RPC) column Shim-pack GIST C18-AQ were purchased from Shimadzu Corporation (Kyoto, Japan). Freund's complete adjuvant (FCA), Freund's incomplete adjuvant (FIA), o-phenylenediamine dihydrochloride (OPD) tablets, 2-mercaptoethanol (2-ME), and all other reagents were obtained from Wako Pure Chemical Industries, Ltd. (Tokyo, Japan) and Wakenyaku Co., Ltd. (Kyoto, Japan).

2.2. Sample Preparation

SO and CHA powders were individually dissolved in 10% DMSO and used in subsequent experiments. The prepared SO and CHA solutions were diluted in 10 mM phosphate-buffered saline (PBS, pH 7.4), serum, or urine and were used as assay samples. In addition, extracts from potatoes were also used as samples. Potatoes (Irish Cobbler, total $n = 12$) were purchased from a supermarket and kept under fluorescent light (380–400 nm wavelength) at 22 °C during the experimental period. Their components were extracted immediately after purchase (Day 0, $n = 4$), after 30 days (Day 30, $n = 4$), and after 60 days (Day 60, $n = 4$). The extracted parts were sprouts (Day 30 and 60), peels (Day 0, 30, and 60), and tubers (Day 0, 30, and 60). Each part was removed from the potatoes (in particular, tubers were carefully removed within 0.5 cm from the peels), weighed, and mixed with 5 mL of 10% DMSO in 15 mL polycarbonate tubes. The parts were homogenized using a 150 Homogenizer (Fisher Scientific Co LLC, Waltham, MA, USA) in the tubes and centrifuged at 1191× *g* for 30 min at 4 °C. These supernatants were used as the assay samples and stored at −80 °C until use. The sample preparation protocol is summarized in Figure 1.

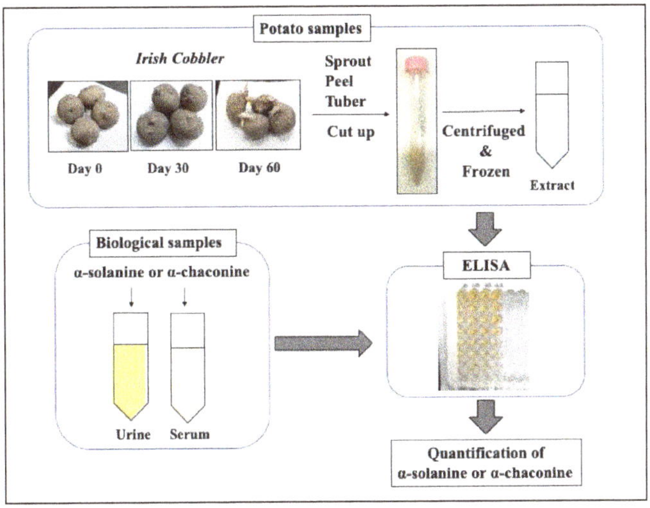

Figure 1. Preparation procedure for each sample. The upper figure shows the preparation process for the potato extract samples. The lower left illustration indicates the preparation of biological samples. Enzyme-linked immunosorbent assays (ELISAs) were conducted using these samples as shown in the lower right image.

2.3. *Development of Solanidine-Binding Antibodies*

The development of antibodies that bind to SO and CHA was completely outsourced to Carbuncle BioScienTec LLC (Kyoto, Japan). This experiment complied with the ARRIVE guidelines (Animal Research: Reporting of In Vivo Experiments) and was approved by the Animal Experiment Committee in Carbuncle BioScienTec LLC. Briefly, solanidine, a chemical compound found in both SO and CHA, was conjugated with bovine serum albumin as an immunogen. Two female rabbits (Japanese White; Nippon Institute for Biological Science; weight, 3.0 kg) were immunized intradermally with 0.5 mg of this immunogen (per rabbit) five times (first time: with FCA, second–fifth time: with FIA) every two weeks. During this period, blood was collected from the ear vein one week after immunization. To obtain the serum, all of these blood samples were left at room temperature (22 °C) for 30 min and then centrifuged at 1191× *g* for 10 min, after which the supernatant was collected. In accordance with a previous report [17], antibody titers in these sera were confirmed by direct ELISA. For the present study, we ensured the collection of sufficient antibody titers, as the results of ELISA performed with 40,000-fold diluted serum after 5 times immunization had previously shown absorbance values of 1.5–2.0 at 450 nm/630 nm wavelengths. Subsequently, blood was drawn from the carotid artery of the rabbits under anesthesia, and 72 mL antiserum was obtained from one rabbit and 69 mL antiserum from the other. Then, rabbit anti-solanidine polyclonal IgGs (anti-sold1 and anti-sold2) were purified from each of the two antisera using an agarose affinity column on which the solanidine-ovalbumin complex was adhered to. The protein concentration of the purified antibodies was quantified using the Lowry method [18]. The procedure for developing these antibodies is summarized in Figure 2. The chemical structures of these compounds were obtained from the PubChem homepage and illustrated using the Ketcher 2.7 software (EPAM Systems Inc. Newtown, PA, USA) [19,20].

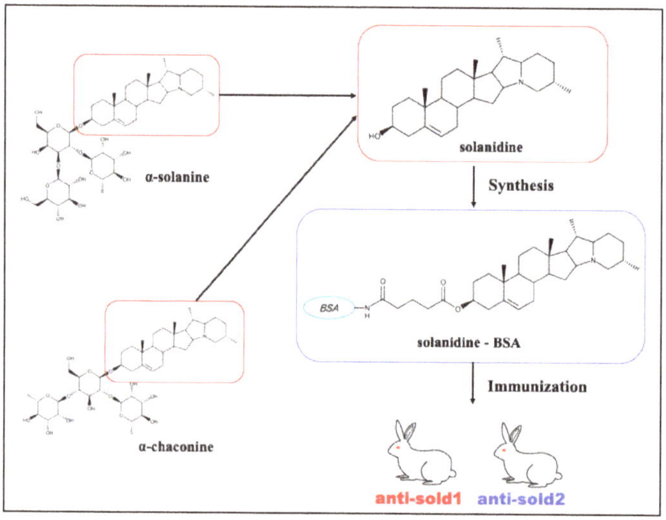

Figure 2. The development procedure for developing solanidine antibodies. The two illustrations on the left show the molecular structures of α-solanine and α-chaconine. These two toxins have the chemical structure of solanidine (the red box on the right). By immunizing two rabbits with a complex of solanidine conjugated with bovine serum albumin (BSA) (solanidine-BSA, the blue box on the right) as the immunogen, two kinds of polyclonal antibodies (anti-sold1 and anti-sold2) were obtained.

2.4. SDS-PAGE and Western Blotting

The purified antibodies were detected by sodium dodecyl sulfate-polyacrylamide gel electrophoresis (SDS-PAGE) as described in a previous study [21]. The concentration of all polyacrylamide gels was 12.5%. After SDS-PAGE, proteins were transferred to nitrocellulose membranes using the Trans-Blot Turbo system (Bio-Rad Laboratories, Inc., Hercules, CA, USA). After blocking the membranes with Blocking One, they were incubated at 4 °C for 1 h with 2 μg/mL HRP-labeled goat anti-rabbit polyclonal IgG (H + L). The membranes were then washed for 5 min thrice with 10 mM Tris-HCl buffer (pH 7.4) and 0.9% NaCl (buffer A), twice with buffer A/0.1% Tween 20, and once with buffer A. Subsequently, antibody-bound proteins were detected using a Chemi-Doc^{TM} XRS plus imaging system and Clarity Western ECL substrate (Bio-Rad Laboratories, Inc.).

2.5. Construction of Non-Competitive Direct ELISAs

Prior to ELISA construction, anti-sold1, and anti-sold2 were labeled to HRP (anti-sold1-HRP and anti-sold2-HRP, respectively) using an HRP labeling kit–SH, according to the manufacturer's protocol. In accordance with a previous report [22], two kinds of non-competitive direct ELISAs were constructed using anti-Sold1-HRP (Sold1 ELISA) and anti-Sold2-HRP (Sold2 ELISA). Briefly, 100 μL of 75 mM carbonate buffer (pH 9.6) was added to each well of a 96-well ELISA Plate H, followed by the addition of 25 μL of sample solutions to the plate and then incubating the plate overnight at 4 °C. After the plate had been washed three times with 10 mM PBS (pH 7.4) and 0.05% Tween 20 (PBS-T), 200 μL of Blocking One was added to each well, and then the plate was incubated for 1 h at 22 °C. The plate was washed with PBS-T again three times, and 100 µL of 2 µg/mL anti-Sold1-HRP or anti-Sold2-HRP diluted with Blocking One was added to each well, followed by incubating the plate for 2 h at 22 °C. Prior to the next plate washing, a chromogenic solution was prepared by dissolving a 5 mg OPD tablet in 10 mL of 0.1 M citrate buffer (pH 5.0). After the plate had been washed three times with PBS-T, 5 µL of 30% H_2O_2 was immediately added to the 10 mL chromogenic solutions, and then 100 µL of the mixture solution was added to each well. After 30 min of color development, the reaction was stopped by adding 100 µL of 3 N H_2SO_4 to each well. The absorbance of the color reaction was measured at a wavelength of 490 nm using a microplate reader (Bio-Rad Laboratories, Inc.). A schematic diagram of this experiment is illustrated in Figure 3.

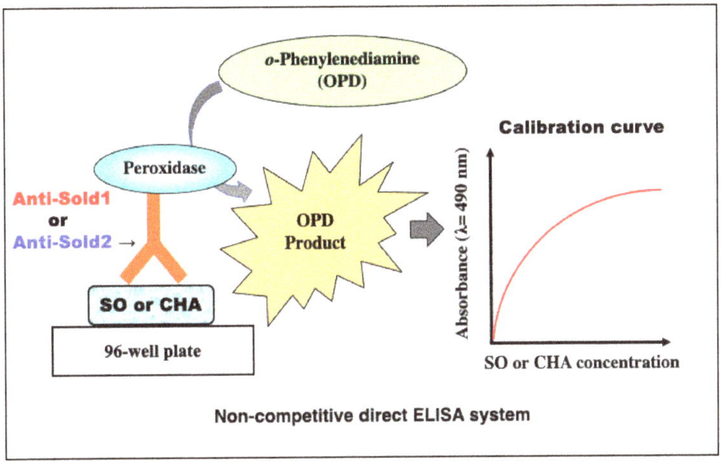

Figure 3. The design of a non-competitive direct ELISA. The left illustration shows the ELISA design used in this study. In this ELISA, the relationship between α-solanine (SO) or α-chaconine (CHA) concentration and the absorbance is denoted by a sigmoid curve, as depicted by the graph on the right.

2.6. HPLC Analysis

In this study, HPLC analysis was performed using the Prominence LC-2010 HPLC system connected to a Shim-pack GIST C18-AQ column based on RPC (Shimadzu Corporation, Kyoto, Japan). Analyses were performed at a room temperature of 22 °C. The flow rate was fixed at 1.0 mL/min, and 10 mM PBS (pH 7.4) was used as the mobile phase. The volume of all samples used for analysis was 100 µL. After sample injection into the HPLC system, a waveform at a wavelength of 210 nm was recorded with a retention time of 30 min. After measurement, the detected waveforms were analyzed using the LabSolutions software version 5.92 provided with the instrument. The lower areas of the waveforms of SO and CHA were calculated using this software to create a calibration curve and to measure the content of SO and CHA in the samples.

2.7. Statistical Analysis

All experiments except for the ones that verified the reproducibility of the ELISAs were measured in quintuplicate, and the data are shown as the mean ± standard deviation (SD). The coefficient of variation (CV) was calculated according to the following equation: (SD/mean) × 100%. In accordance with a previous report [23], the LOD was defined as the mean concentration of blank samples (n = 16) + 3SD. The recovery of SO and CHA in each ELISA was calculated according to the following equation: recovery = [(final concentration − initial concentration)/added concentration] × 100%. To create the calibration curve, this formula and Spearman's rank correlation coefficient (R-value) were calculated using the KaleidaGraph software version 5.0.4 (Hulinks Inc., Tokyo, Japan).

3. Results

3.1. Detection of Purified Polyclonal Antibodies

To confirm that the rabbit polyclonal antibodies were successfully purified, SDS-PAGE and Western blotting were performed. Although the serum contained many components (Figure 4A,B, and lane 1), only IgGs were purified by eluting the proteins bound to the affinity column (Figure 4A,B and lane 2). To confirm that the purified protein is IgG, the fragmentation was verified by adding 2-ME to the sample. In the sample without 2-ME, the bands of IgG had a molecular weight of 150 kDa (Figure 4C,D and lanes 3–7); however, in the sample with 2-ME, the bands disappeared, and fragmentation was confirmed by SDS-PAGE and Western blotting (Figure 4C,D and lanes 1–2). We succeeded in purifying anti-sold1 and anti-sold2 and thus proceeded to construct the ELISAs.

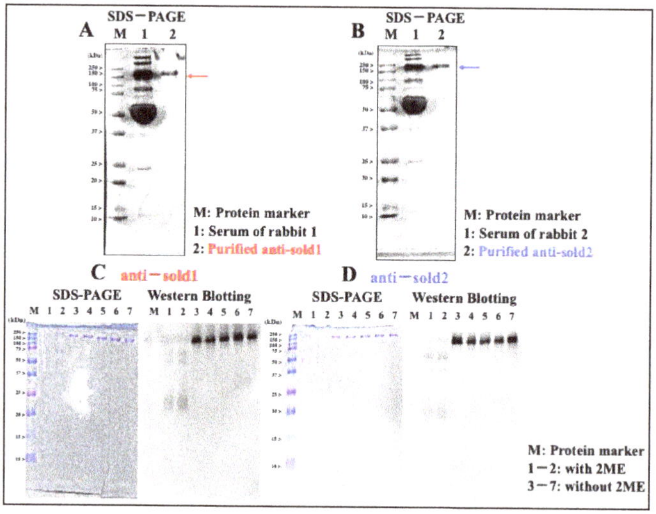

Figure 4. The detection of purified antibodies. (**A**,**B**) The results of sodium dodecyl sulfate-polyacrylamide gel electrophoresis (SDS-PAGE) showing the serum components and antibodies. Lanes 1 and 2 show the bands of serum components and purified antibodies, respectively, (red and blue arrows) obtained from two different rabbits. (**C**,**D**) The results of SDS-PAGE (left panel) and Western blotting (right panel) showing the bands of antibodies with/without 2-mercaptoethanol (2-ME). Lanes 1 and 2 show the bands of purified antibodies with 2-ME obtained from the two different rabbits. Lanes 3–7 show the bands of purified antibodies without 2-ME obtained from the two different rabbits. Lane M in all the pictures indicates a protein marker, and the numbers on the left side of the lane represent the molecular weight (kDa) of each component. The SDS-PAGE results depicted in panels C and D were photographed with a digital camera, and those depicted in the other panels were captured using a ChemiDoc^{TM} XRS plus imaging system (Bio-Rad Laboratories Inc.).

3.2. Construction of Two ELISAs to Detect SO and CHA

Two non-competitive direct ELISAs were constructed using anti-sold1 and anti-sold2. When the seven two-fold serial dilutions with PBS were measured, Sold1 ELISA showed good calibration curves for SO and CHA in the concentration range of 1.56–100 ng/mL (Figure 5A,C), whereas Sold2 ELISA was able to make calibration curves in the concentration range of 3.12–100 ng/mL (Figure 5B,D). The calculated LODs of SO had better values in the Sold1 ELISA (1.38 ng/mL) than in the Sold2 ELISA (2.95 ng/mL) (Tables 1 and 2). Similarly, the LODs of CHA were better in the Sold1 ELISA (1.08 ng/mL) than in the Sold2 ELISA (2.76 ng/mL) (Tables 1 and 2). Furthermore, the detection sensitivity of CHA was slightly higher than that of SO in both of the ELISAs.

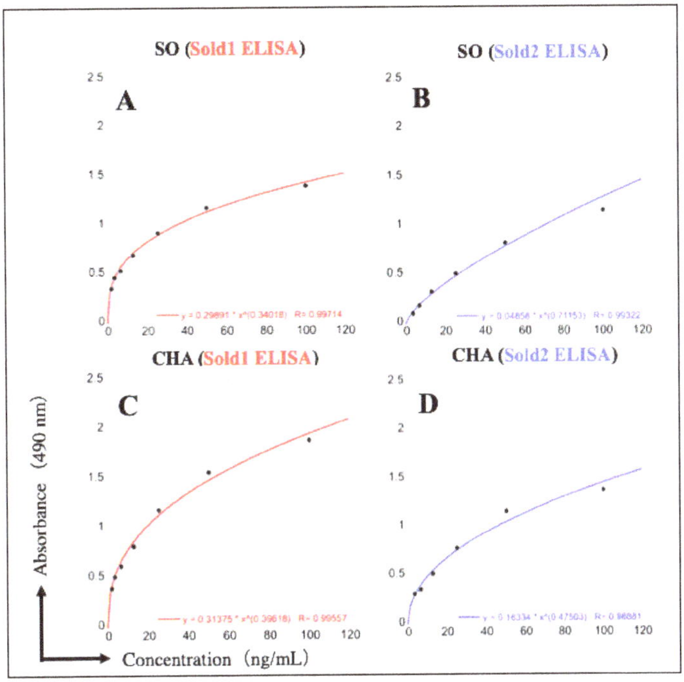

Figure 5. The calibration curves of each ELISA using serial dilutions of the SO and CHA in 10 mm phosphate-buffered saline (PBS, pH 7.4). The blank-subtracted absorbance mean (black circles) ± SD (error bars) values were plotted against SO (**A**,**B**) and CHA (**C**,**D**) concentrations to construct the calibration curves of Sold1 ELISA (**A**,**C**) and Sold2 ELISA (**B**,**D**). In each graph, the *x*-axis indicates the SO or CHA concentrations (ng/mL) and the *y*-axis indicates their absorbance at 490 nm. The samples for all ELISAs were seven two-fold serial dilutions of the SO and CHA standard (1.56–100 ng/mL), using PBS as the diluent and blank. The calibration curves were graphed, and the formula and R-value were calculated using the KaleidaGraph software.

Table 1. The detection performance of Sold1 ELISA based on its calibration curves obtained using PBS serial dilutions.

	Sold1 ELISA	100 (ng/mL)	50 (ng/mL)	10 (ng/mL)		Sold1 ELISA	100 (ng/mL)	50 (ng/mL)	10 (ng/mL)
SO in PBS	Mean (ng/mL)	98.49	51.74	10.89	CHA in PBS	Mean (ng/mL)	101.70	52.76	9.64
	SD	5.58	4.50	1.72		SD	5.62	3.45	1.32
	CV (%)	5.66	8.70	15.80		CV (%)	5.53	6.55	13.71
	Recovery (%)	97.46	95.20	89.20		Recovery (%)	105.28	101.58	98.20
	LOD (ng/mL)		1.38			LOD (ng/mL)		1.08	
SO in serum	Mean (ng/mL)	196.95	102.38		CHA in serum	Mean (ng/mL)	208.83	114.44	
	SD	16.85	14.30			SD	15.29	14.27	
	CV (%)	8.55	13.96			CV (%)	7.32	12.47	
	Recovery (%)	122.18	119.82			Recovery (%)	120.52	126.84	
SO in urine	Mean (ng/mL)	60.73	22.16		CHA in urine	Mean (ng/mL)	78.53	39.85	
	SD	5.14	3.77			SD	4.66	5.08	
	CV (%)	8.47	17.03			CV (%)	5.93	12.74	
	Recovery (%)	78.26	68.56			Recovery (%)	84.18	81.65	

Sixteen equivalent samples were used for each measurement condition.

Table 2. The detection performance of Sold2 ELISA based on its calibration curves obtained using PBS serial dilutions.

	Sold2 ELISA	100 (ng/mL)	50 (ng/mL)	10 (ng/mL)		Sold2 ELISA	100 (ng/mL)	50 (ng/mL)	10 (ng/mL)
SO in PBS	Mean (ng/mL)	101.01	49.23	8.14	CHA in PBS	Mean (ng/mL)	100.89	51.45	9.67
	SD	4.88	4.55	2.36		SD	3.63	4.22	1.88
	CV (%)	4.83	9.24	28.96		CV (%)	3.60	8.21	19.48
	Recovery (%)	102.25	97.68	90.18		Recovery (%)	101.95	104.25	96.42
	LOD (ng/mL)		2.95			LOD (ng/mL)		2.76	
SO in serum	Mean (ng/mL)	146.98	84.44		CHA in serum	Mean (ng/mL)	162.35	90.88	
	SD	12.63	13.26			SD	12.47	11.70	
	CV (%)	8.59	15.71			CV (%)	7.68	12.88	
	Recovery (%)	114.56	125.82			Recovery (%)	120.88	129.27	
SO in urine	Mean (ng/mL)	50.79	19.11		CHA in urine	Mean (ng/mL)	57.01	26.87	
	SD	4.54	3.65			SD	4.47	4.76	
	CV (%)	8.95	19.11			CV (%)	7.83	17.71	
	Recovery (%)	77.25	68.28			Recovery (%)	78.82	77.16	

Sixteen equivalent samples were used for each measurement condition.

3.3. *Verification of the Repeatability of Sold1 ELISA*

To evaluate the repeatability of Sold1 ELISA, 16 samples each of SO and CHA diluted to 10 ng/mL, 50 ng/mL, and 100 ng/mL with PBS, serum, and urine as solvents were measured. When the SO and CHA samples diluted in PBS were measured, Sold1 ELISA showed excellent CV and simultaneous repeatability (Figure 6A,D and Table 1). However, a non-specific chromogenic reaction was observed in the samples in which SO and CHA were diluted with serum, and the 10 ng/mL samples could not be measured, whereas the calculated concentrations of the 50 ng/mL and 100 ng/mL samples were approximately 2-fold higher than the actual concentrations (Figure 6B,E and Table 1). Then, the same serum samples were deproteinized by adding 10% trichloroacetic acid, after which they were measured by ELISA. Although the non-specific reaction was suppressed by the deproteinization treatment of the serum, the coloration itself was hardly observed. Furthermore, the chromogenic reaction was suppressed in the samples in which SO and CHA were diluted in urine. The 10 ng/mL samples could not be measured, whereas the 50 ng/mL and 100 ng/mL samples showed somewhat lower values than their actual concentrations (Figure 6C,F and Table 1).

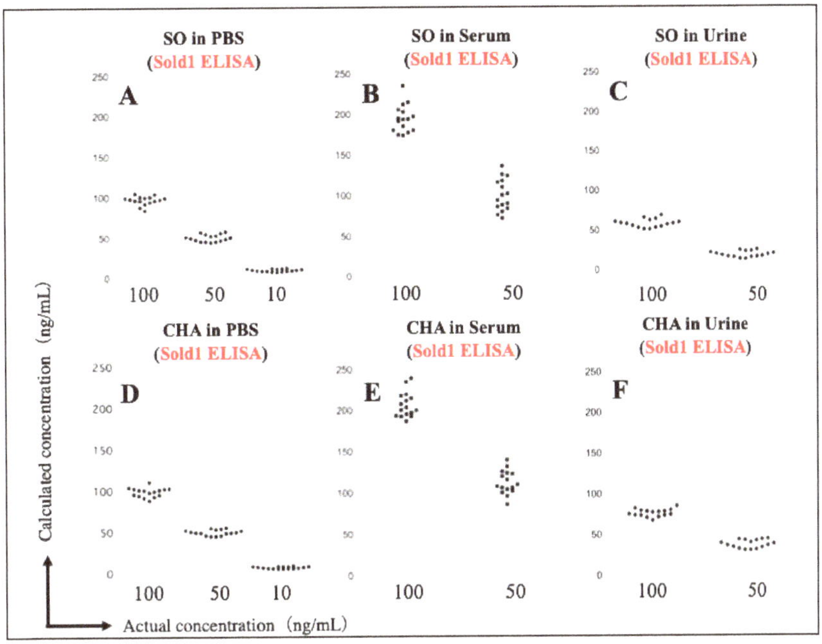

Figure 6. The repeatability of Sold1 ELISA for measuring SO and CHA concentrations in different biological matrices. The Sold1 ELISA was used to simultaneously measure 16 samples of SO (**A–C**) and CHA (**D–F**), each at concentrations of 10 ng/mL (PBS only), 50 ng/mL, and 100 ng/mL in different biological matrices (PBS: circles, serum: squares, and urine: diamonds). Columns indicate the detection of SO and CHA in PBS (**A**,**D**), serum (**B**,**E**), and urine (**C**,**F**). In the graphs, the *x*-axis indicates the actual concentrations (ng/mL) of the samples prepared by dissolving SO and CHA powders in these matrices and the *y*-axis indicates the SO and CHA concentrations (ng/mL) calculated from the Sold1 ELISA calibration curves (Figure 5A,C). The dotted lines indicate the average of the measurements.

3.4. Verification of the Repeatability of Sold2 ELISA

The repeatability of Sold2 ELISA was investigated using the same procedure as for Sold1 ELISA. The concentrations of SO and CHA diluted in PBS were accurately measured with only small variations (Figure 7A,D and Table 2). Slightly nonspecific chromogenic reactions were observed in the samples of SO and CHA diluted with serum, with 10 ng/mL not being measurable and 50 ng/mL and 100 ng/mL calculated to concentrations about 1.5 times higher than the actual concentrations (Figure 7B,E and Table 2). The chromogenic reaction was inhibited in the samples of SO and CHA diluted in urine, with 10 ng/mL not being measurable and 50 ng/mL and 100 ng/mL calculated to be about half the actual concentration (Figure 7C,F and Table 2).

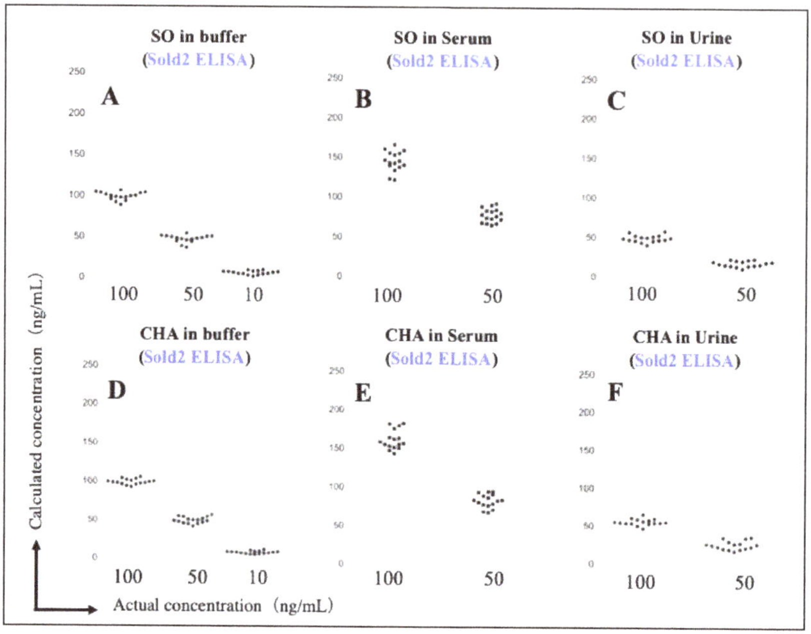

Figure 7. The repeatability of Sold2 ELISA for measuring the SO and CHA concentrations in different biological matrices. Sold2 ELISA was used for measuring 16 equivalent SO (**A–C**) and CHA (**D–F**) samples with concentrations of 10 ng/mL (PBS only), 50 ng/mL, and 100 ng/mL in different biological matrices (PBS: circles, serum: squares, and urine: diamonds). Columns indicate the detection of SO and CHA in PBS (**A**,**D**), serum (**B**,**E**), and urine (**C**,**F**). In the graphs, the x-axis indicates the actual concentrations (ng/mL) of the samples prepared by dissolving SO and CHA powders in these matrices and the y-axis indicates the SO and CHA concentration (ng/mL) calculated by Sold2 ELISA calibration curves (Figure 5B,D). The dotted lines indicate the average of the measurements.

3.5. Reconstruction of Sold1 ELISA Calibration Curves Using the Dilution Series of SO and CHA in Serum and Urine

To correctly measure the SO and CHA concentrations in the serum and urine, Sold1 ELISA calibration curves were newly constructed from the measurement results of the seven two-fold dilution series (12.5–800 ng/mL) using serum and urine as their solvents. Good calibration curves of SO and CHA diluted in serum could be made in the concentration range of 25–800 ng/mL (Figure 8A,B). The LOD of SO in serum was 15.25 ng/mL, whereas that of CHA in serum was 13.48 ng/mL (Table 3). Meanwhile, calibration curves of SO and CHA diluted in urine could be constructed in the concentration range of 50–800 ng/mL (Figure 8C,D). The LOD of SO in urine was 30.28 ng/mL, whereas that of CHA in urine was 27.92 ng/mL (Table 3). Based on these calibration curves, 16 samples of 100 ng/mL SO and CHA diluted in serum were re-measured, and the simultaneous repeatability was improved (Figure 8E and Table 3). The repeatability of simultaneous measurements was also improved when 16 samples of 100 ng/mL SO and CHA diluted in urine were re-measured (Figure 8F and Table 3).

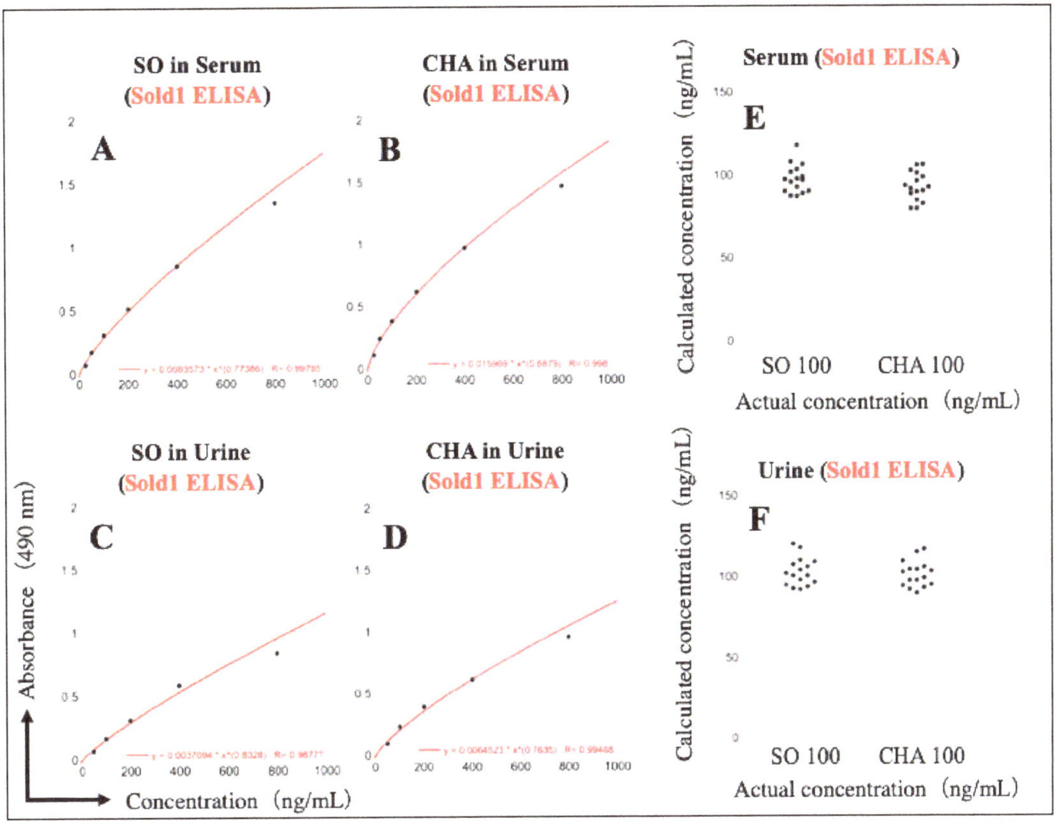

Figure 8. Reconstruction of the calibration curves to improve the accuracy of SO and CHA quantification in serum and urine by Sold1 ELISA. To adjust for the interfering substances present in serum and urine, new calibration curves were constructed for serum (**A**,**B**) and urine (**C**,**D**), based on seven serial two-fold dilutions of SO and CHA (12.5–800 ng/mL), using each biological matrix type as the diluent and blank. The blank-subtracted absorbance mean (black circles) ± SD (error bars) values were plotted against SO (**A**,**C**) and CHA (**B**,**D**) concentrations to construct the calibration curves by Sold1 ELISA. In each graph, the x-axis indicates the SO and CHA concentrations (ng/mL) and the y-axis indicates their absorbance at 490 nm. The calibration curves were graphed, and the formula and R-value were calculated using the KaleidaGraph software. The repeatability of SO and CHA quantification in serum (**E**) and urine (**F**) was re-evaluated using the new calibration curves, via 16 simultaneous measurements of SO and CHA in these biological matrices (serum: squares, urine: diamonds). In the graphs, the x-axis indicates the actual concentrations (100 ng/mL) prepared by the SO and CHA powders diluted in these matrices, and the y-axis indicates the SO and CHA concentrations (ng/mL) calculated by Sold1 ELISA reconstructed calibration curves (**A**–**D**). The dotted lines indicate the average of the measurements.

Table 3. The detection performance of ELISAs based on the corrected calibration curves obtained using the serum and urine serial dilution measurements.

	Sold1 ELISA	100 (ng/mL)	Sold2 ELISA	100 (ng/mL)
SO in serum	Mean (ng/mL)	98.48	Mean (ng/mL)	96.55
	SD	8.42	SD	8.62
	CV (%)	8.55	CV (%)	8.92
	Recovery (%)	96.58	Recovery (%)	94.15
	LOD (ng/mL)	15.25	LOD (ng/mL)	19.41
SO in urine	Mean (ng/mL)	93.59	Mean (ng/mL)	102.20
	SD	8.65	SD	9.43
	CV (%)	9.24	CV (%)	9.23
	Recovery (%)	90.88	Recovery (%)	105.64
	LOD (ng/mL)	30.28	LOD (ng/mL)	45.16
CHA in serum	Mean (ng/mL)	103.60	Mean (ng/mL)	98.97
	SD	8.50	SD	9.29
	CV (%)	8.20	CV (%)	9.38
	Recovery (%)	106.62	Recovery (%)	96.67
	LOD (ng/mL)	13.48	LOD (ng/mL)	16.92
CHA in urine	Mean (ng/mL)	102.02	Mean (ng/mL)	103.69
	SD	7.97	SD	8.22
	CV (%)	7.81	CV (%)	7.93
	Recovery (%)	105.74	Recovery (%)	106.58
	LOD (ng/mL)	27.92	LOD (ng/mL)	38.15

3.6. Reconstruction of Sold2 ELISA Calibration Curves Using the Dilution Series of SO and CHA in Serum and Urine

Similarly, the Sold2 ELISA calibration curves for serum and urine were re-created and verified for simultaneous repeatability. Sold2 ELISA was able to construct calibration curves for SO and CHA diluted in serum in the concentration range of 25–800 ng/mL (Figure 9A,B). The LOD of SO in serum was 19.41 ng/mL, while that of CHA in serum was 16.92 ng/mL (Table 3). Further, the urinary SO and CHA could be measured in the concentration range of 50–800 ng/mL (Figure 9C,D). The LOD of SO in urine was 45.16 ng/mL, whereas that of CHA in urine was 38.15 ng/mL (Table 3). Based on these calibration curves, the simultaneous measurement of 16 samples of 100 ng/mL SO and CHA diluted in serum and urine greatly improved (Figure 9D,E and Table 3).

3.7. Construction of Calibration Curve for Simultaneous Measurement of SO and CHA

We examined whether it is possible to simultaneously measure both SO and CHA in each solvent. Good calibration curves could be drawn for SO and CHA samples diluted in PBS in the concentration range of 1.56–100 ng/mL in Sold1 ELISA and 3.12–100 ng/mL in Sold2 ELISA (Figure 10A,D). However, for the samples containing both SO and CHA diluted in serum, calibration curves could be constructed in the concentration range of 25–800 ng/mL in the two ELISAs (Figure 10B,E). Similarly, for urine samples containing both SO and CHA, calibration curves could be made in the concentration range of 50–800 ng/mL in the two ELISAs (Figure 10C,F).

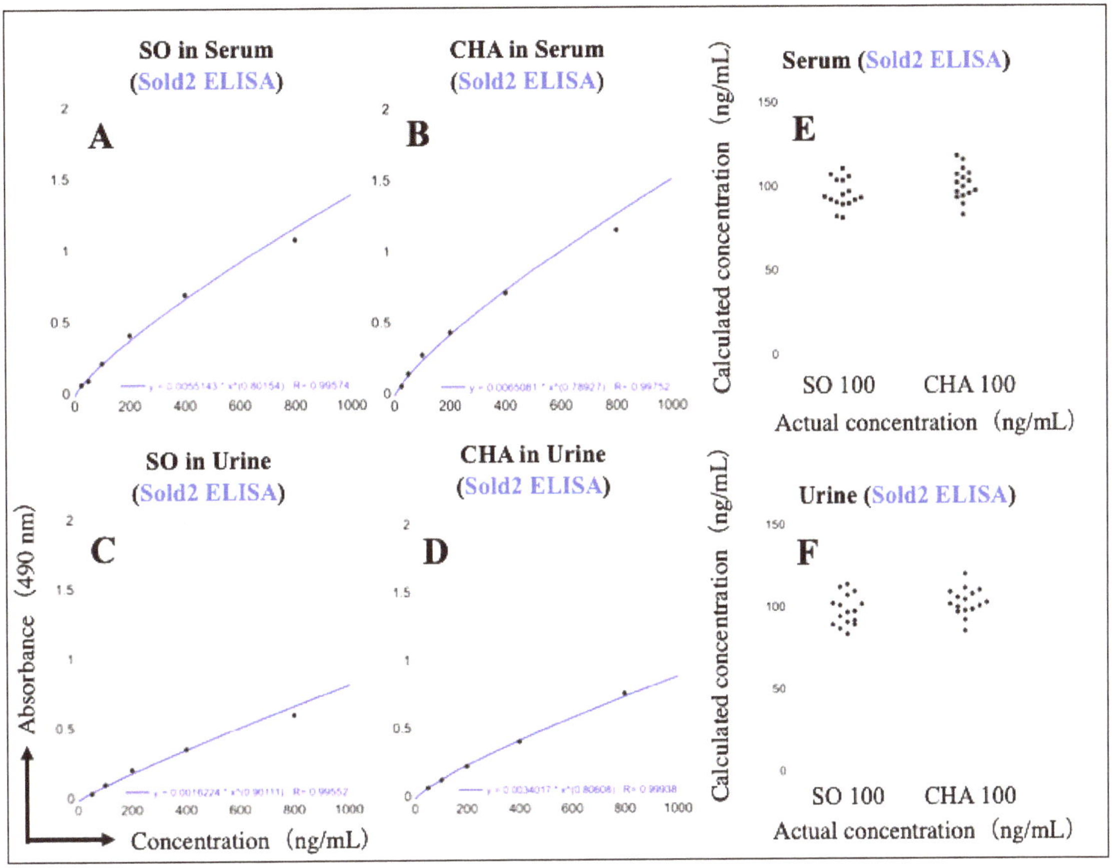

Figure 9. Reconstruction of calibration curves to improve the accuracy of SO and CHA quantification in serum and urine by Sold2 ELISA. To adjust for the interfering substances present in serum and urine, new calibration curves were constructed for serum (**A**,**B**) and urine (**C**,**D**), based on seven serial two-fold dilutions of SO and CHA (12.5–800 ng/mL), using each biological matrix type as the diluent and blank. The blank-subtracted absorbance mean (black circles) ± SD (error bars) values were plotted against SO (**A**,**C**) and CHA (**B**,**D**) concentrations to construct calibration curves by Sold2 ELISA. In each graph, the x-axis indicates the SO and CHA concentration (ng/mL) and the y-axis indicates their absorbance at 490 nm. The calibration curves were graphed, and the formula and R-value were calculated using the KaleidaGraph software. The repeatability of SO and CHA quantification in serum (**E**) and urine (**F**) was re-evaluated using the new calibration curves, via 16 simultaneous measurements of SO and CHA in these biological matrices (serum: squares, urine: diamonds). In the graph, the x-axis indicates the actual concentrations (100 ng/mL) prepared by diluting SO and CHA powders in these matrices, and the y-axis indicates the SO and CHA concentrations (ng/mL) calculated by Sold2 ELISA reconstructed calibration curves (**A**–**D**). The dotted lines indicate the average of the measurements.

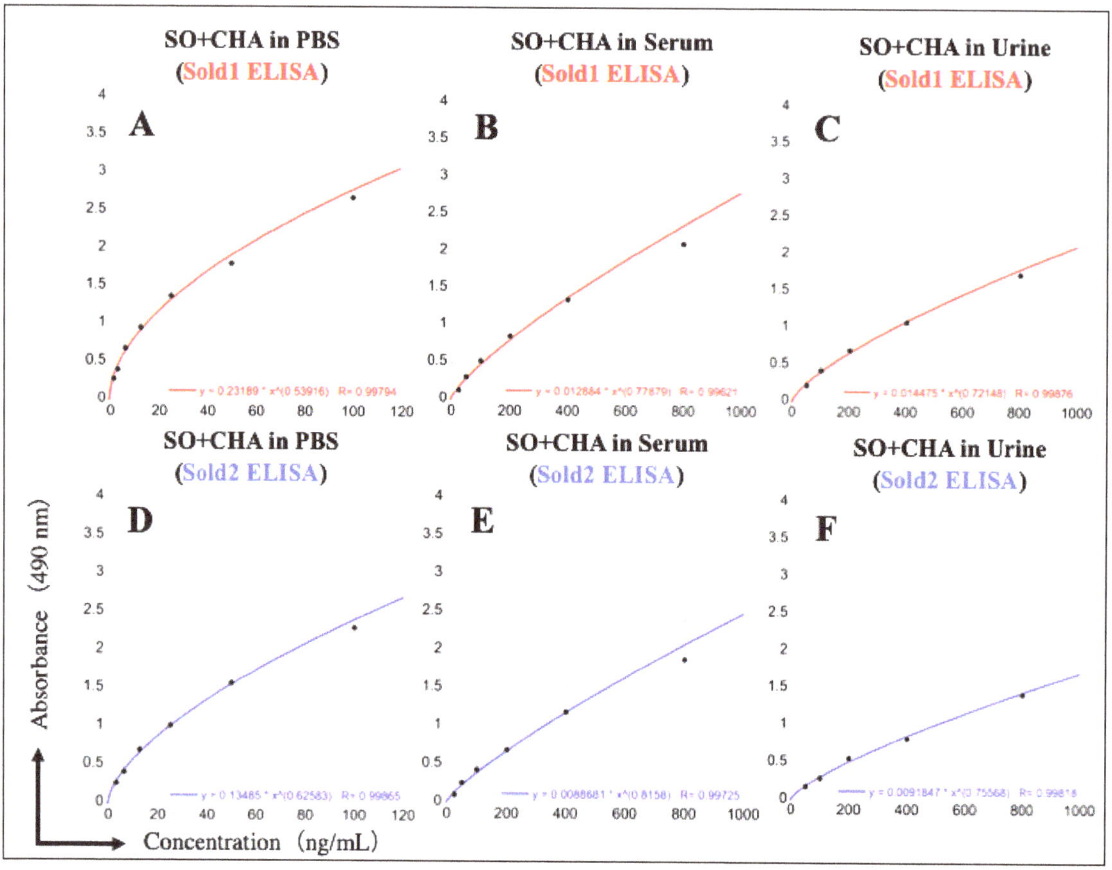

Figure 10. Calibration curves of each ELISA using serial dilutions with equal concentrations of the SO and CHA mixture in each matrix. The blank-subtracted absorbance mean (black circles) ± SD (error bars) values were plotted against SO and CHA mixture concentration to construct calibration curves by Sold1 ELISA (**A–C**) and Sold2 ELISA (**D–F**). In each graph, the *x*-axis indicates the SO and CHA concentration (ng/mL) in each mixture (e.g., "100" on the *x*-axis means that both the SO and CHA concentrations were 100 ng/mL in the mixture), whereas the *y*-axis indicates absorbance at 490 nm. The samples for all ELISAs were seven two-fold serial dilutions of the SO and CHA mixture standard, using PBS (1.56–100 ng/mL, **A**,**D**), serum (12.5–800 ng/mL, **B**,**E**), and urine (12.5–800 ng/mL, **C**,**F**) as the diluent and blank. The calibration curves were graphed, and the formula and R-value were calculated using the KaleidaGraph software. SO + CHA; SO and CHA were in equal concentrations in the mixture.

3.8. Detection Performance of SO and CHA in Potato Extracts

The concentration of SO and CHA in each part of the potato was determined using ELISAs. In potato tubers, the SO and CHA content increased about 1.6- to 1.8-fold from Day 0 to Day 60 (Figure 11A,D). SO and CHA was detected in the peel extracts at about 180 ng/mL on Day 0 and increased to more than two-fold (~400 ng/mL) on Day 60 (Figure 11B,E). However, SO and CHA were detected in the extract of sprouts at about 3000 ng/mL on Day 30, and the same or higher levels of both chemicals were found on Day 60 (Figure 11C,F). In addition, no color development was observed in the samples prepared with distilled water at concentrations of 800 ng/mL of starch, pectin, glucose, and fructose, which are nutrients that are abundant in potatoes.

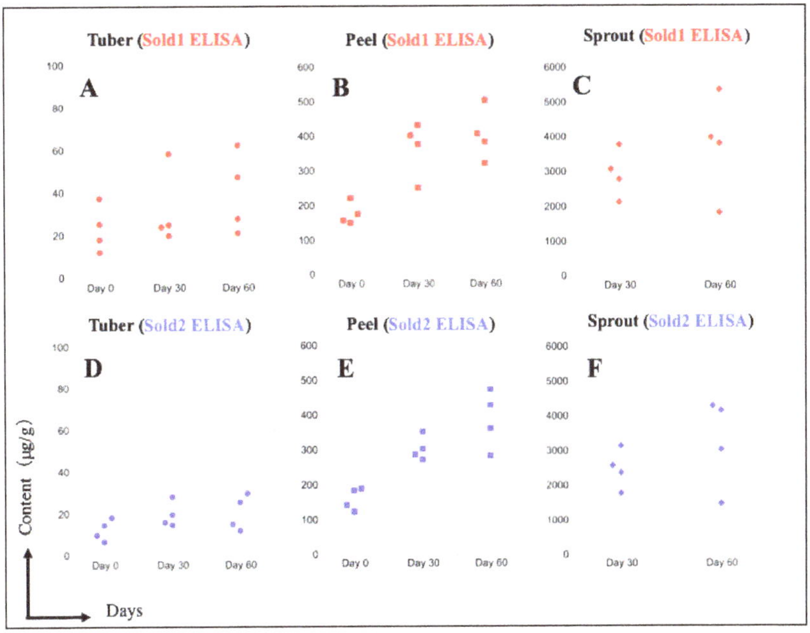

Figure 11. Quantification of SO and CHA concentrations in potato extracts using ELISA. Irish Cobbler potatoes (n = 4 on each Day) were used to prepare the extracts of each potato part. After measuring the concentrations of SO and CHA in each part extract, based on calibration curves constructed by PBS serial dilutions (Figure 10A,D), the content was calculated by multiplying their weights. The SO and CHA concentrations (μg/g) in tuber (**A**,**D**, circle), peel (**B**,**E**, square), and sprout (**C**,**F**, diamond), as calculated by Sold1 ELISA (**A**–**C**, red) and Sold2 ELISA (**D**–**F**, blue). In each graph, the x-axis indicates the elapsed period since potato purchase (Day 0, 30, and 60) and the y-axis indicates the content in each part (μg/g). The dotted lines indicate the average of the measurements.

3.9. HPLC Analysis of SO and CHA in Potato Extracts

Finally, the concentration of SO and CHA in each part of the potato was analyzed by HPLC. When the mixture of SO and CHA dissolved in PBS was analyzed, their main waveforms were detected at a retention time of 4–6 min (Figure 12A). By using the relationship between the concentrations of SO and CHA in PBS and the area under the waveform at this retention time, a good calibration curve could be drawn in the concentration range of 12.5–800 ng/mL (Figure 12B). However, when the extracts from each potato part were measured with HPLC, no clear waveforms could be detected using a retention time of 4–6 min (Figure 12C–E). Therefore, the waveforms that were barely detected during this time period were used to calculate the SO and CHA contents of each potato part. Differences in the results between the ELISA and HPLC analyses of the SO and CHA calculated content for the same potato (n = 4) parts on Day 60 are shown in Table 4.

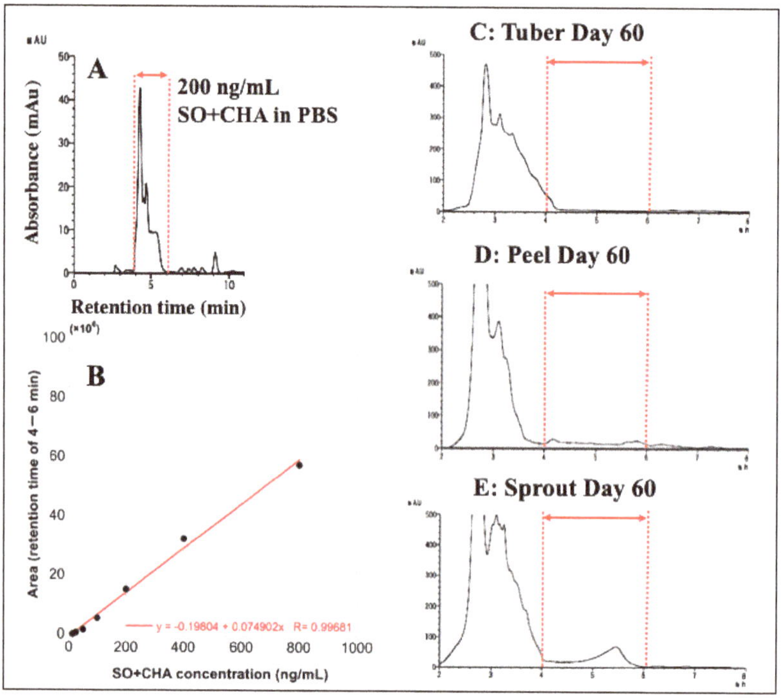

Figure 12. HPLC analysis of SO and CHA in PBS and potato extracts. A 200 ng/mL SO and CHA mixture diluted in PBS were analyzed by HPLC (**A**). The *x*-axis shows the retention time (min), and the *y*-axis shows the absorbance (mAu) at 210 nm. The waveforms at a retention time of 4–6 min (surrounded by red dotted lines) were determined to reflect mainly SO and CHA in solution. A calibration curve of HPLC using serial dilutions of equal concentrations of SO and CHA mixture in PBS (**B**). The calculated area's mean (black circles) ± SD (error bars) values were plotted against SO and CHA mixture concentration to construct calibration curves by HPLC. The *x*-axis indicates the SO and CHA concentrations (ng/mL) in PBS, whereas the *y*-axis indicates the calculated areas under the waveform at a retention time of 4–6 min. The samples for HPLC analysis were 7 two-fold serial dilutions of the SO and CHA mixture standard using PBS (12.5–800 ng/mL). The calibration curve was graphed, and the formula and R-value were calculated using the KaleidaGraph software. The tuber (**C**), peel (**D**), and sprout (**E**) extracts derived from Irish Cobbler potatoes on Day 60 were analyzed by this HPLC. The *x*-axis shows the retention time (min), and the *y*-axis shows absorbance (mAu) at 210 nm. Retention times of 4–6 min, where SO and CHA are expected to be mainly detected, are delimited by red dotted lines. The area under the waveform detected in this range was matched with the calibration curve (**B**), and the calculated SO and CHA contents (µg/g) are shown in Table 4. SO + CHA, SO, and CHA are in equal concentrations in the mixture.

Table 4. Differences between ELISA and HPLC analyses of the SO and CHA calculated content of the same potato parts on Day 60.

		Calculated Content (μg/g) Using Sold1 ELISA		
	(n = 4)	**Tuber**	**Peel**	**Sprout**
Potato on Day 60	**No.1**	22.04	389.0	1892
	No.2	63.28	510.2	3868
	No.3	48.20	327.1	5423
	No.4	28.68	412.4	4038
		Calculated Content (μg/g) Using Sold2 ELISA		
	(n = 4)	**Tuber**	**Peel**	**Sprout**
Potato on Day 60	**No.1**	15.68	365.1	1538
	No.2	30.28	478.4	3102
	No.3	26.22	288.0	4348
	No.4	12.25	432.8	4212
		Calculated Content (μg/g) Using HPLC		
	(n = 4)	**Tuber**	**Peel**	**Sprout**
Potato on Day 60	**No.1**	8.27	170.3	1065
	No.2	13.95	209.8	1982
	No.3	10.16	152.7	2643
	No.4	8.66	202.0	2129

4. Discussion

Here, we successfully developed two new polyclonal antibodies that can bind to both SO and CHA and constructed two ELISA systems based on these antibodies. Although these antibodies are currently owned only by us, they can be produced by any institute by immunizing rabbits with the solanidine-BSA complex shown in Figure 2. Moreover, the immunization of mice with this solanidine-BSA complex may allow for the creation of monoclonal antibodies with better sensitivity and specificity for the detection of SO and CHA. Thus, our results provide essential information for all scientists involved in potato food poisoning research.

As potatoes contain both SO and CHA, developing antibodies that can bind to both SO and CHA would be the best way to construct an ELISA to detect them. Therefore, we aimed to establish antibodies that can recognize a part of solanidine, a chemical structure shared by SO and CHA, as epitopes. However, the molecular weight of solanidine is considerably low (400 Da), and the epitopes to which antibodies can bind are extremely limited. Since anti-sold1 and anti-sold2 in this study probably recognized almost the same epitope, we determined that it would be difficult to construct an analytical method using two different antibodies that recognize different epitopes, such as a sandwich ELISA system. Thus, we established non-competitive direct ELISA systems using anti-sold1 and anti-sold2. Both the ELISAs in our study were more sensitive to CHA than to SO, and the Sold1 ELISA appeared to have better detection performance than the Sold2 ELISA.

The constructed ELISAs in this study are comparable to the existing assays in terms of their ability to detect SO and CHA diluted in buffer solution. Our ELISA systems are about 20 times more sensitive in terms of LODs than previous ELISA kits that use monoclonal antibodies [10]. In particular, the LODs measuring SO and CHA in PBS by Sold1 ELISA are comparable to the results of previous HPLC analyses [8,11]. Although the Sold1 ELISA is not as sensitive as the combination of HPLC and mass spectrometry for SO and CHA [12], our ELISAs were good assays in terms of accuracy and precision, at least for the measurement of these chemicals diluted in PBS. Unfortunately, the sensitivity of our ELISA for SO and CHA in serum was low. A previous study reported an HPLC assay system that measured GAs (SO and CHA) in serum at a LOD of 0.3 ng/mL [13], which is about 15 times more sensitive than our ELISAs. In the same report, the peak serum concentrations after the ingestion of SO at 0.41 mg/kg body weight and CHA at 0.59 mg/kg body weight were 8 ng/mL and 14 ng/mL, respectively [13], which cannot be measured by our ELISAs. Thus, the detection performance of SO and CHA in serum needs to be improved in the future. In our experiment, when serum was used as the matrix, the blank containing only serum without SO or CHA showed higher absorbance, which may have contributed to the overall low detection sensitivity. This non-specific reaction may be caused by the strong influence of several components in the serum, especially the protein component. In fact, the addition of 10% trichloroacetic acid to the serum suppressed nonspecific coloration, suggesting that the deproteinization in serum samples is very important. Because of the great variety of proteins in serum, it would be difficult to narrow down the protein components that cross-react with antibodies, so an efficient protein removal method is desired. However, since the strong deproteinization also seemed to remove SO and CHA from the samples, a more careful examination is needed to optimize this process. A previous report suggested that immunoglobulins in the serum sporadically bind to each well of an ELISA plate and that the binding of HRP-labeled antibodies to these immunoglobulins causes false-positive reactions [24]. To solve this problem, the deproteinization of serum samples appears to be a good approach. However, the method of protein removal by denaturing protein components in serum with acids or organic solvents may affect SO and CHA. Since the protein removal method based on filtration is easy to apply, it is considered an ideal pre-treatment method for ELISA. Thus, there are many options for deproteinization, each of which has its own merits and demerits that cannot be verified in this study, but we would like to challenge them in the future. Because there are few previous reports on the measurement of SO and CHA in urine samples, comparing the detection sensitivity of our method with that of existing methods was impossible. However, in the present study, the LODs of SO and CHA were inferior in urine than in serum, and we must consider that the detection performance in urine was poor. When ELISA was performed, the coloration of urine samples was weaker than that of the PBS and serum samples, and the absorbance of the urine samples was lower than that of the other samples. This is probably due to the high concentration of urea in the urine samples. Since urea is known to act as a protein-denaturing agent [25], we speculated that the urea bound to each well of the ELISA plate denatured the HRP-labeled antibodies. Since a method to completely remove only urea from urine has not yet been established, we expect that it would be difficult to improve this detection performance of SO and CHA in urine. For both urine and serum, it would be desirable to prepare affinity columns conjugated with anti-sold1 or anti-sold2 and treat the samples with them to specifically separate SO and CHA from the biological components. Considering the CV and recovery, the variation and accuracy of SO and CHA diluted in serum or urine in ELISA are inferior to those of SO and CHA diluted in PBS. At present, it would be difficult to adapt our ELISA as a laboratory method for patients with potato food poisoning, but with further improvement, it may be feasible.

Although the Sold1 ELISA and Sold2 ELISA values were slightly different, the combined concentrations of SO and CHA could be measured from the extracts of each part of the potato. The SO and CHA concentrations in most tubers of commercial potatoes do not exceed 100 μg/g [26], whereas those in sprouts and peels are in the ranges of 2000–4000 μg/g and 300–600 μg/g, respectively [27]. Our experimental results were in the range of these concentrations or slightly higher. Each part of the potato contains not only SO and CHA but also solanidine and other GAs, and antibodies can bind to these chemicals as well, resulting in rather high values. In addition, it is also important to verify the cross-reactivity of antibodies with the components in potatoes other than GAs. Basically, 100 g of potato contains a very high amount of carbohydrates (21.44 g), with some fiber (2.3 g) and sugars (1.08 g) [28]. Our ELISAs did not react with starch (carbohydrates), pectin (fiber), and monosaccharides (sugars), suggesting that these assays effectively captured the toxic components in the potatoes. However, the accuracy of the results of the HPLC analysis is doubtful, since no clear waveforms indicating SO and CHA could be obtained from the potato extract samples. Since the contents of SO and CHA were underestimated by this HPLC analysis, this does not seem to be appropriate as a food testing method. It is not clear whether ELISA or HPLC analysis is more accurate, but given the unstable detection waveform of HPLC, ELISA seemed to have an advantage. As anti-sold1 and anti-sold2 bind to the substances with solanidine structure, it is necessary to further investigate the cross-reactivity of these polyclonal antibodies in detail. Although it would not reach the level of food poisoning, eating potatoes with a SO and CHA content exceeding 140 μg/g gives a bitter taste, and eating potatoes with SO and CHA content exceeding 220 μg/g contributes to a burning sensation in the mouth and throat [4,26]. To adopt our ELISAs as food testing methods, they need to be accurate enough to distinguish these subtle differences in SO and CHA concentration. We would like to re-examine each step of our ELISAs and develop a more accurate assay. Furthermore, it is important to re-establish competitive ELISA and indirect ELISA methods to enhance detection performance. In competitive ELISA, the antigens are pre-labeled, and samples containing a large amount of components other than the target chemicals can also be used [29]. As competitive ELISA requires only one type of antibody, it is relatively inexpensive and has been widely adopted in medical facilities. In contrast, the indirect ELISA method is cost-prohibitive as it uses a labeled secondary antibody in addition to the primary antibody, but it is expected to significantly improve the detection sensitivity [29]. In the future, we will attempt to construct ELISA methods based on these different principles so that researchers can customize the assay for different purposes and situations.

Non-competitive direct ELISAs are advantageous because several samples can be measured simultaneously on a single ELISA plate. In this respect, ELISA is superior to HPLC because the analysis of one sample must be completed before the analysis of the next sample can begin in the HPLC method. Another advantage of ELISA is that it does not require expensive equipment and is relatively easier to perform. ELISA only requires a microplate reader, which is less expensive than HPLC or mass spectrometry instruments. Moreover, optimizing anti-sold1 and anti-sold2 as medical reagents will enable their adaption to immunological analyzers that are already installed in medical facilities. Another merit of ELISA is the easy analysis of results—they can be interpreted visually as the concentrations of SO and CHA are indicated by the degree of coloration. The interpretation of HPLC and mass spectrometry results requires a certain degree of skills and expertise, and their use in medical facilities is limited. Notably, most of the previous analytical methods for SO and CHA were reported before 2000 AD [8–10,13], and there has been little development of detection methods for these two toxins in the last 20 years. Under such circumstances, our proposed method of creating polyclonal antibodies that bind to SO and CHA, and ELISAs that use these antibodies, possesses a high novelty. Even if only the two antibodies produced in this study are used, it is possible to construct new indirect ELISAs, sandwich ELISAs, and competitive ELISAs; moreover, further development would be expected as well. This increase in options for measuring SO and CHA is academically important and will provide useful evidence for many researchers involved in laboratory medicine and food testing.

ELISA also has some limitations, the primary one being that the measurement time currently takes more than one day. The first step of overnighting the samples on the plate is time-consuming, and a future challenge would be to shorten the measurement time. The second limitation is the high running cost. ELISA requires the use of HRP-conjugated antibodies for each assay, which are costlier than HPLC reagents. Additionally, detection sensitivity is a weak point when serum or urine is used as samples.

While there have been limited assay methods for SO and CHA, the development of non-competitive ELISAs is a useful achievement for the future development of potato food poisoning diagnosis and food testing. In the future, we aim to seek optimal conditions for ELISA with the goals of shortening the analysis time and improving sensitivity and specificity. Furthermore, we would like to start the construction of an immunochromatography using the antibodies developed in this study to establish a more rapid and simple method for measuring SO and CHA.

5. Conclusions

In this study, non-competitive direct ELISAs that can measure SO and CHA in both serum and urine samples were developed. However, the detection sensitivity remains an issue. Although cross-reactions with other components in the biological samples may occur, these ELISAs can measure SO and CHA in potato extracts. Therefore, these ELISAs would not be applicable to clinical testing but may be useful for food testing. The improvement of the sample preparation process and assay conditions can increase the applicability of these ELISAs in the future.

Author Contributions: K.O. contributed to project administration, funding acquisition, resources, conceptualization, methodology, investigation, data curation, formal analysis, and writing of the original draft. K.M. contributed to formal analysis, validation, visualization, writing, review, and editing. All authors have read and agreed to the published version of the manuscript.

Funding: This work was supported by MHLW KA Program Grant Number JPMH 21KA3007.

Data Availability Statement: The data presented in this study are available in this article.

Acknowledgments: We would like to thank Kazuyuki Sawadaishi and Chiwa Kataoka in Carbuncle BioScienTec LLC for developing antibodies in this research.

Conflicts of Interest: The authors declare no conflict of interest.

References

1. Friedman, M. Potato Glycoalkaloids and Metabolites: Roles in the Plant and in the Diet. *Mol. Nutr. Food Res.* **2007**, *51*, 116–134. [CrossRef] [PubMed]
2. Slanina, P. Solanine (glycoalkaloids) in potatoes—toxicological evaluation. *Food Chem. Toxicol.* **1990**, *28*, 759–761. [CrossRef] [PubMed]
3. Friedman, M.; McDonald, G.M. Potato glycoalkaloids: Chemistry, analysis, safety, and plant physiology. *Crit. Rev. Plant Sci.* **1997**, *16*, 55–132. [CrossRef]
4. Sinden, S.L.; Deahl, K.L.; Aulenbach, B.B. Effect of glycoalkaloids and phenolics on potato flavor. *J. Food Sci.* **1976**, *41*, 520–523. [CrossRef]
5. Bushway, R.J.; Rathy, P. a-Chaconine and a-Solanine Content of Potato Products and Their Stability during Several Modes of Cooking. *J. Agric. Food Chem.* **1981**, *29*, 814–817. [CrossRef]
6. Wilson, A.M.; McGann, D.F.; Bushway, R.J. Effect of Growth-Location and Length of Storage on Glycoalkaloid Content of Roadside-Stand Potatoes as Stored by Consumers. *J. Food Prot.* **1983**, *46*, 119–121. [CrossRef]
7. Phillips, B.J.; Hughes, J.A.; Phillips, J.C.; Walters, D.G.; Anderson, D.; Tahourdin, C.S. A study of the toxic hazard that might be associated with the consumption of green potato tops. *Food Chem. Toxicol.* **1996**, *34*, 439–448. [CrossRef]
8. Sotelo, A.; Serrano, B. High-performance liquid chromatographic determination of the glycoalkaloids α-solanine and α-chaconine in 12 commercial varieties of Mexican potato. *J. Agric. Food Chem.* **2000**, *48*, 2472–2475. [CrossRef]
9. Stanker, L.H.; Kamps-Holtzapple, C.; Friedman, M. Development and characterization of monoclonal antibodies that differentiate between potato and tomato glycoalkaloids andaglycons. *J. Agric. Food Chem.* **1994**, *42*, 2360–2366. [CrossRef]
10. Friedman, M.; Bautista, F.F.; Stanker, L.H.; Larkin, K.A. Analysis of Potato Glycoalkaloids by a New ELISA Kit. *J. Agric. Food Chem.* **1998**, *46*, 5097–5102. [CrossRef]
11. Kodamatani, H.; Saito, K.; Niina, N.; Yamazaki, S.; Tanaka, Y. Simple and sensitive method for determination of glycoalkaloids in potato tubers by high-performance liquid chromatography with chemiluminescence detection. *J. Chromatogr. A* **2005**, *1100*, 26–31. [CrossRef] [PubMed]
12. Popova, I.; Sell, B.; Pillai, S.S.; Kuhl, J.; Dandurand, L.M. High-Performance Liquid Chromatography–Mass Spectrometry Analysis of Glycoalkaloids from Underexploited Solanum Species and Their Acetylcholinesterase Inhibition Activity. *Plants* **2022**, *11*, 269. [CrossRef] [PubMed]
13. Hellenäs, K.E.; Nyman, A.; Slanina, P.; Lööf, L.; Gabrielsson, J. Determination of potato glycoalkaloids and their aglycone in blood serum by high-performance liquid chromatography. Application to pharmacokinetic studies in humans. *J. Chromatogr.* **1992**, *573*, 69–78. [CrossRef] [PubMed]
14. Mensinga, T.T.; Sips, A.J.; Rompelberg, C.J.; van Twillert, K.; Meulenbelt, J.; van den Top, H.J.; van Egmond, H.P. Potato glycoalkaloids and adverse effects in humans: An ascending dose study. *Regul. Toxicol. Pharmacol.* **2005**, *41*, 66–72. [CrossRef] [PubMed]
15. Hansen, A.A. Two fatal cases of potato poisoning. *Science* **1925**, *61*, 348–349. [CrossRef]

16. McMillan, M.; Thompson, J.C. An outbreak of suspected solanine poisoning in school boys: An examination of criteria of solanine poisoning. *Q. J. Med.* **1979**, *48*, 227–243.
17. Stephens, D.B.; Thomas, R.E.; Stanton, J.F.; Iverson, B.L. Polyclonal antibody catalytic variability. *Biochem. J.* **1998**, *332*, 127–134. [CrossRef]
18. Lowry, O.H.; Rosebrough, N.J.; Farr, A.L.; Randall, R.J. Protein measurement with the Folin phenol reagent. *J. Biol. Chem.* **1951**, *193*, 265–275. [CrossRef]
19. PubChem Homepage. Available online: https://pubchem.ncbi.nlm.nih.gov (accessed on 30 March 2023).
20. Epam/Ketcher Homepage. Available online: https://github.com/epam/ketcher/blob/master/LICENSE (accessed on 30 March 2023).
21. Towbin, H.; Staehelin, T.; Gordon, J. Electrophoretic transfer of proteins from polyacrylamide gels to nitrocellulose sheets: Procedure and some applications. *Proc. Natl. Acad. Sci. USA* **1979**, *76*, 4350–4354. [CrossRef]
22. Koike, A.; Arai, S.; Yamada, S.; Nagae, A.; Saita, N.; Itoh, H.; Uemoto, S.; Totani, M.; Ikemoto, M. Dynamic mobility of immunological cells expressing S100A8 and S100A9 in vivo: A variety of functional roles of the two proteins as regulators in acute inflammatory reaction. *Inflammation* **2012**, *35*, 409–419. [CrossRef]
23. Lim, M.; Erdman, P.; Cho, S.; Mathew, A.; Fleisher, M.; Thoren, K.L. Evaluation of CisBio ELISA for Chromogranin A Measurement. *J. Appl. Lab. Med.* **2019**, *4*, 11–18. [CrossRef] [PubMed]
24. Terato, K.; Do, C.T.; Cutler, D.; Waritani, T.; Shionoya, H. Preventing intense false positive and negative reactions attributed to the principle of ELISA to re-investigate antibody studies in autoimmune diseases. *J. Immunol. Methods* **2014**, *407*, 15–25. [CrossRef] [PubMed]
25. Song, M.F.; Li, Y.S.; Ootsuyama, Y.; Kasai, H.; Kawai, K.; Ohta, M.; Eguchi, Y.; Yamato, H.; Matsumoto, Y.; Yoshida, R.; et al. Urea, the most abundant component in urine, cross-reacts with a commercial 8-OH-dG ELISA kit and contributes to overestimation of urinary 8-OH-dG. *Free. Radic. Biol. Med.* **2009**, *47*, 41–46. [CrossRef] [PubMed]
26. Barceloux, D.G. Potatoes, Tomatoes, and Solanine Toxicity (*Solanum tuberosum* L., *Solanum lycopersicum* L.). *Dis. Mon.* **2009**, *55*, 391–402. [CrossRef]
27. Lachman, J.; Hamouz, K.; Orsak, M.; Pivec, V. Potato glycoalkaloids and their significance in plant protection and human nutrition-Review. *Rostl. Vyroba.* **2001**, *47*, 181–191.
28. McGill, C.R.; Kurilich, A.C.; Davignon, J. The role of potatoes and potato components in cardiometabolic health: A review. *Ann. Med.* **2013**, *45*, 467–473. [CrossRef]
29. Tabatabaei, M.S.; Ahmed, M. Enzyme-Linked Immunosorbent Assay (ELISA). *Methods Mol. Biol.* **2022**, *2508*, 115–134.

Disclaimer/Publisher's Note: The statements, opinions and data contained in all publications are solely those of the individual author(s) and contributor(s) and not of MDPI and/or the editor(s). MDPI and/or the editor(s) disclaim responsibility for any injury to people or property resulting from any ideas, methods, instructions or products referred to in the content.

MDPI
St. Alban-Anlage 66
4052 Basel
Switzerland
www.mdpi.com

Foods Editorial Office
E-mail: foods@mdpi.com
www.mdpi.com/journal/foods

Disclaimer/Publisher's Note: The statements, opinions and data contained in all publications are solely those of the individual author(s) and contributor(s) and not of MDPI and/or the editor(s). MDPI and/or the editor(s) disclaim responsibility for any injury to people or property resulting from any ideas, methods, instructions or products referred to in the content.

www.ingramcontent.com/pod-product-compliance
Lightning Source LLC
Chambersburg PA
CBHW061625210926
51850CB00009B/2443
* 9 7 8 3 0 3 6 5 8 5 6 2 8 *